豪门
贵族府邸典藏

The Aristocrats　Classic Noble Mansion

ID Book 图书工作室 编

华中科技大学出版社
http://www.hustp.com
中国·武汉

PREFACE

Using Design Contain the Whole Life

Chinese people have a persistent pursuit on residence which stems from a permanent and sustainable adore. Since ancient times, the study of the residence has pushed the development of the residential culture.

An ideal residence not only can satisfy daily need but also is a symbol of quality life. In some people's mind, luxury villa has much space to waste, while in fact it is not true. Nowadays, luxury villa is bearing more club function except for being a complete private place. With the innovation of technology and development of times, the spirit of the past can be restored in ancient ways, meanwhile the space can be more full of concept of "Home". Louis-Ambroise Dubut once vividly described: "Are there any place more comfortable than one's home? It brings you happiness; it lets you live in happiness every day."

Past and present, simple and sophisticated, classic and current, different voices fiercely collided with each other. It seems like the Big Bang giving birth to many new spaces. In fact, design is like life-By drinking water, one knows hot and cold. Complicated consideration often fall behind the free voice in people's mind. If one has the spirit of independence and the idea of freedom, then any place can be his house. Hundreds of design is an expression of material civilization. For someone with design ideal, he can show things from the deep memory and express the attitude towards the world.

Giving up the thinking of "The so-called luxury villa" while getting into the nature of life pursuit. I am keeping on telling my group that design comes from life, only real life leads to the origin of design.

Pang Yifei
Top-Ten Influential Designer from 2012 to 2013 of China International Architectural Decoration and Design Expo
President of Chongqing Pinchen Decoration Construction Design Co., Ltd.
High-Class Legalized Designer of APDC International Design Communication Center
Chief of Joint Meeting of CIDEA

The basement centers on leisure and entertainment, adjusting the location of audio-visual room, making full use of the lighting of sunken courtyard and making the basement become more spacious and brighter, with functions of table tennis area, audio-visual area, bar area, wine cellar area, cigar area, chess and card room and spa fitness area.

这是一套面积达942.8 m²的英式豪宅。

一层主要以客厅、餐厅、起居室为主要功能空间，还设置了一间套房作为父母房。

早餐厅、中餐厅和西餐厅围绕着在西式厨房周边，让三个空间互动起来。

在客厅区域设置了一个空间作为起居室，既让功能得到满足，也使整体空间更为宽敞、气派。

二层在总体平面布局的基础上，对主卧的更衣室作了调整，使主卧空间更为完整、开阔。对主卫也作了调整，使空间更为奢华。

地下室主要以休闲娱乐功能为主，调整了视听室的位置，充分利用下沉式庭院的采光，让地下室的空间更为宽敞、明亮，功能设置为桌球区、视听区、酒吧区、酒窖区、雪茄区、棋牌室和水疗健身区。

THE ARISTOCRATS CLASSIC NOBLE MANSION

THE ARISTOCRATS CLASSIC NOBLE MANSION

豪门 贵族府邸典藏

THE ARISTOCRATS CLASSIC NOBLE MANSION

THE ARISTOCRATS CLASSIC NOBLE MANSION

Silver Bay Chinese Style Private Mansion

银海湾中式私宅

Design Company: Space Impression Decorative Design
Project Location: Zhuhai of Guangdong Province
Project Area: 600 m²

设计公司：空间印象建筑装饰设计
项目地点：广东珠海
项目面积：600 ㎡

Home is the place to inherit aesthetics.

"The significance of life does not lie in what we have, but in what we can leave for the world." Similarly, "the significance of a mansion is in what we can teach or pass on to the residents, yet the focus of education can not live without respect for traditions and explorations for the future". The designer compares architecture in the past with modern architecture, and integrates classical architectural aesthetics into modern and concise life. And in design, the details prevail.

The mansion is like the objects, and the mansion is the man. In the conceptual proposal for this project, the designer firstly presents luxury watch and swimming pool presenting life quality, demonstrating the magnificence and high-level of this project. At the same time, the exquisite space decorations can set off the uncommon temperament of the property owner, creating harmonious space environment for the noble residents, leading to the concept of "mansion is the man".

家，是美学的传承

"人生的意义不在于拥有什么，而在于能给世间留下什么。"同理，"豪宅的意义在于我们可以将什么传达给居住者，而教育的中心，则离不开对传统的尊重，对未来的探索"。设计师将过去的建筑和现代的建筑进行了比较，将古典建筑美学融入到现代的简约生活中，在设计上以细节取胜。

宅如其物，宅如其人。在本项目的概念提案上，设计师首先展示的是奢侈品名表及代表生活品质的游泳池，通过"宅如其物"来彰显项目的高端、大气。同时，用精致的空间装饰品来陪衬主人的不凡气质，为尊贵的居住者打造和谐的空间环境，引申出了"宅如其人"的概念。

THE ARISTOCRATS CLASSIC NOBLE MANSION

THE ARISTOCRATS CLASSIC NOBLE MANSION

Hong Kong Jiulongcang Sorrento Example Flat

香港九龙仓·擎天半岛示范单位

Design Company: Hong Kong Fong Wong Architects & Associates
Designer: Noah Fong
Design Team: Fong Wong (Shenzhen)
Project Location: Chengdu of Sichuan Province
Project Area: 228 m²
Major Materials: Local Environmental Materials
Photographer: Ye Jingxing

设计公司：香港方黄建筑师事务所
设 计 师：方峻
设计团队：方黄（深圳）
项目地点：四川成都
项目面积：228 m²
主要材料：当地环保材料
摄 影 师：叶景星

Not "Tuhao", graceful reflection of gold.
Gold color is a word that can make people think of any nobility and elegance.

In the main space of cold color tones such as white, gray and bluish purple, the glittering gold light is like a highlighting pen, making the whole space be dynamic, and full of texture.

非土豪，金的曼妙辉映
金色能让人想起任何一个高贵典雅的词汇。

在白色、灰色、蓝紫等冷色调为主体的空间里，抢眼的金色光影好似一支点睛笔，将空间瞬间烘托得生动活泼，且富有质感。

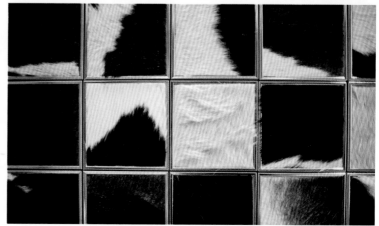

THE ARISTOCRATS CLASSIC NOBLE MANSION

To Define French Romance · Oak Bay Villa

◆ 定义法式浪漫·橡树湾别墅 ◆

Design Company: Chongqing Pinchen Decorative Design and Engineering Co., Ltd.
Designers: Pang Yifei, Li Jian, Yin Zhengyi
Soft Decoration Designer: Cheng Jing
Project Location: Chongqing
Project Area: 412 m²
Major Materials: Stone, Pattern Paint, Gold Outlining, Hand-painted Oil Painting, Solid Wood Floor

设计公司：重庆品辰装饰设计工程有限公司
硬装设计师：庞一飞、李健、殷正毅
软装设计师：程静
项目地点：重庆
项目面积：412 ㎡
主要材料：石材、做旧显纹漆、线条描金处理、手绘油画、实木地板等

Living inside this space of French romantic atmosphere, what we feel are the space colors of French romance. Exotic charms and warm life are presented in front of our eyes, making us know how to enjoy life and enjoy the moment.

Capturing the instant of artistic inspirations and combining abstract with reality, the dynamic design is integrated into the space. Upon walking into the hall, you can find that the high space displays the detail aesthetic feel of the design is displayed to the extremes. The harmonious symmetric feel, the gilding picture and the curve edge of home furnishings, the ups and downs of various lines in the tranquil space make us seem to go back to the French ancient castle several centuries ago.

The elegant spiral-type staircase, the wooden handrail has fine touch feel, and when sunshine shines through, all become softer. The hand-painted wall, the romantic scarlet sofa, the transparent crystal lights and the pink bed, among the rich colors, the monotonous space format is abandoned, and all is like a romantic century movie.

THE ARISTOCRATS CLASSIC NOBLE MANSION

生活在这个具有法式浪漫气息的空间中，我们感受到的是法兰西浪漫的空间色彩，异国风情和温馨生活呈现在眼前，让我们懂得享受生活、享受当下。

设计师捕捉瞬间的艺术灵感，将抽象与现实相结合，把带有生机的设计融入到空间中。步入大厅，挑高的空间将设计细节上的美感发挥到极致。和谐的对称感、鎏金的描画、陈设家居的曲线边缘，各种线条在静态空间中的起伏，使我们仿佛穿越到数世纪前的法式古堡中。

典雅的螺旋型楼梯，木质扶手触感极好，阳光洒入的时候，变得更加轻柔。手绘的墙面、浪漫的猩红沙发、剔透的水晶灯、粉红的女儿床，在丰富的色彩中，摒弃了单调的空间形态，仿佛是一场浪漫的世纪电影。

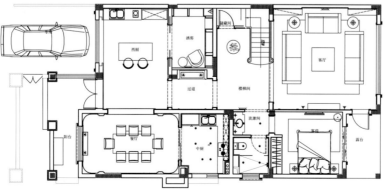

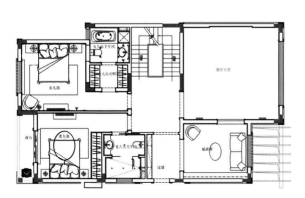

— 043 —

THE ARISTOCRATS CLASSIC NOBLE MANSION

THE ARISTOCRATS CLASSIC NOBLE MANSION

Rongqiao New City Legend Villa Show Flat

融侨新城泷郡别墅样板房

Design Company: Shanghe Design Consultants Co., Ltd.	设计公司：上合设计顾问有限公司
Designer: Yu Zhoulin	设 计 师：余周霖
Associate Designers: Ye Zhiying, Wang Zhou	参与设计：叶志应、王舟
Project Location: Fuqing of Fujian Province	项目地点：福建福清
Project Area: 850 m²	项目面积：850 ㎡
Major Materials: Burl Wood Paint, Leather, Mirror Surface Stainless Steel, Marble, Solid Wood Composite Floor	主要材料：树瘤木烤漆、皮革、镜面不锈钢、大理石、实木复合地板
Photographers: Sanxiang Photography, Zhang Jing	摄　　影：三像摄　张静

The house type for single villa should have the most comfortable and most cozy structure of life space, which shall not just meet with simple family activities such as having meals and sleeping. The multiple functional space divisions enrich people's home residential life. This space has elevators, convenient for the host to move upwards and downwards inside home. Entering home is like stepping into a private little world.

As a public space, the first floor is a place for the host to receive guests. The living room neighbors the dining hall. The living room is as tall as 2 floors. The space is decorated with grand and magnificent crystal drop-lights, which make the space do not appear abrupt at all. The whole French window makes the space appear quite bright and the scenery outside the window naturally becomes the background. The tone of the living room is quite concise, the sofa in circling format becomes the main part of the living room and the

豪门 贵族府邸典藏

water blue color velvet sofa produces the most elegant atmosphere. Stone is the main material for the space. The space is decorated with various formats and various colors of marble. The sedate atmosphere makes the space appear grand and magnificent.

This project is attached with many bedrooms, each bedroom is a tiny suit of complete facilities, collocated with leisure space, changing room and grand washroom. The complete equipment makes the master's life become more convenient and life become more comfortable.

Other than that, the space has wine bar, tea room and entertainment space. It could make do for a group of friends to gather here. With brisk and fashionable modern style, the whole residential space displays the temperament of style residence, making the concept of "residing first" go through the whole space.

独栋别墅结构的户型应该是生活空间最为宽裕、舒适的户型结构，这样的空间不仅仅满足吃饭、睡觉等简单的家庭活动。多样化的功能分区丰富了人们家居生活的乐趣。本案空间还配有电梯，方便主人在家中上下活动，进入家中就好像进入了一个私家的小世界。

一层作为公共空间，是主人招待客人的地方。客厅与餐厅比邻，客厅的挑高高达2层楼，装上华丽硕大的水晶吊灯，也丝毫不觉突兀。整面的落地窗子，让空间显得格外透亮，而窗外的美景也自然地成为背景。客厅的基调十分简约，围绕一圈的沙发就是客厅的主体，水蓝的绒布沙发，显示出优雅的气息。石料是空间的主材，各种模样、颜色的大理石装点空间，沉稳的氛围使空间大气而美观。

本案有多间卧室，每间卧室都是一间设备齐全的小型套房，配备有休憩区、更衣间、大型卫生间等，使主人的生活更加便利、生活更加舒适。

另外，酒吧、茶室、娱乐室一应俱全，邀请一群好友小聚也完全可以满足。整个居室空间以利落、时尚的现代风格演绎了格调家居的气质，使居住至上的理念贯穿始终。

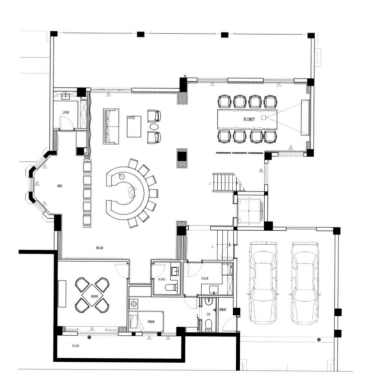
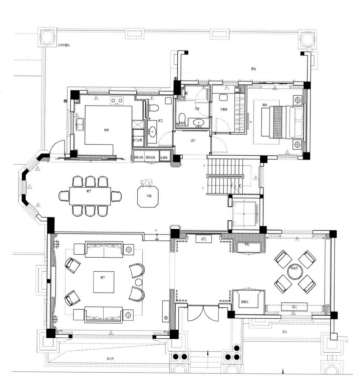

豪门 贵族府邸典藏

THE ARISTOCRATS CLASSIC NOBLE MANSION

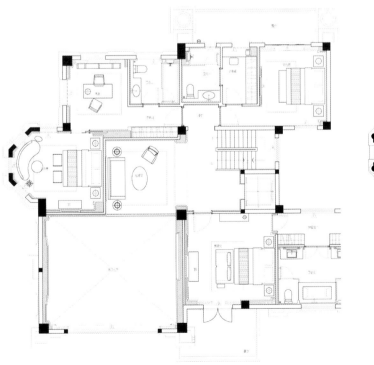
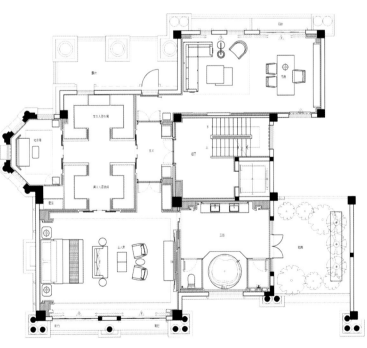

THE ARISTOCRATS CLASSIC NOBLE MANSION

Suzhou Longchi Mansion, Boutique Show Flat

苏州龙池华府精装样板间

Interior Design Company: Hangzhou EHE Interior Design Co., Ltd.
Space Designer: Ma Hui
Soft Decoration Furnishing Company: Hangzhou Jishang Design and Decorative Engineering Co., Ltd.
Layout Designer: Ge Xulian
Project Location: Suzhou
Project Area: 567m²

室内设计公司：杭州易和室内设计有限公司
空间设计师：马辉
软装陈设公司：杭州极尚装饰设计工程有限公司
陈设设计师：葛旭莲
项目地点：江苏苏州
项目面积：567 ㎡

For the project design, the designer tries to present American style life interests which are leisurely, elegant and natural, thus relieving the pressure feel brought by modern life, and displaying a relaxing and free life space.

Compared with other spaces, the living room appears brighter and fresher, creating some sparkling feel for people upon entering the room, and providing a warm and pleasing atmosphere for the communication between host and guests. Inside the whole residence, the solid wood furniture, bronze accessories, iron furniture and perfect color collocations create some refreshing and leisurely physical and psychological sensations for people.

THE ARISTOCRATS CLASSIC NOBLE MANSION

豪门 贵族府邸典藏

THE ARISTOCRATS CLASSIC NOBLE MANSION

traveling around the world. He would select artistic objects from all over the world by himself and enjoy the process.

Hostess: 45-year-old, housewife. She used to be an opera actress and came to Shenzhen with her husband for further development. Three years ago, she quit her job to pay more attention to her daughter, and guided her onto the road of music, fostering the likes for artistic objects. She is especially fond of flower arrangement.

Child: 19-year-old only daughter, major in music, enthusiastic and easy-going. She would like to study abroad.

Life Background

As a successful businessman traveling inside the country and abroad, the host longs for some elegant life just like the musical tones of Vienna while enjoying the joys of success, just like Swarovski crystal, graceful as always. The hostess is a housewife, who would cook some health tea and study floriculture during the leisure time. She quite enjoys communicating with the daughter on music and sometimes she would taste the artistic objects with the host that they select from abroad. Other than that, she would spend a lot of time with her friends for shopping or beautifying, and attend various kinds of parties. As their daughter spends a lot of time at home, thus the family would taste some dessert, communicate and talk about art on the balcony...

豪门 贵族府邸典藏

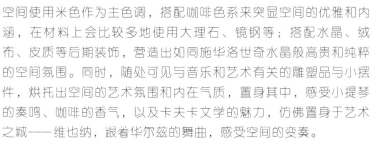

空间使用米色作为主色调，搭配咖啡色系来突显空间的优雅和内涵，在材料上会比较多地使用大理石、镜钢等；搭配水晶、绒布、皮质等后期装饰，营造出如同施华洛世奇水晶般高贵和纯粹的空间氛围。同时，随处可见与音乐和艺术有关的雕塑品与小摆件，烘托出空间的艺术氛围和内在气质，置身其中，感受小提琴的奏鸣、咖啡的香气，以及卡夫卡文学的魅力，仿佛置身于艺术之城——维也纳，跟着华尔兹的舞曲，感受空间的变奏。

家庭成员

男主人： 48岁，进出口贸易商，对世界各地的艺术品十分喜爱。工作中严谨、大气，但是生活中开朗、细心，对生活的品质要求比较高，喜欢周游世界，会亲自去挑选世界各国的艺术品，并乐在其中。

女主人： 45岁，全职太太。曾经是歌剧院演员，后随丈夫来深圳

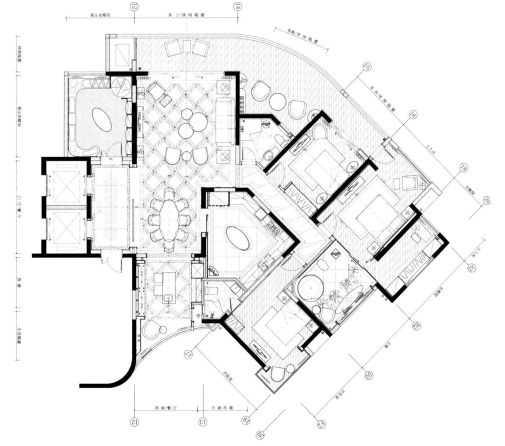

发展，并于三年前放弃工作，专心陪伴女儿，引导她走上音乐之路，喜欢艺术品，并对插花情有独钟。

小孩：19岁的独生女就读音乐专业，性格开朗、热情，将来打算出国深造。

生活背景

男主人作为一个经常往返于国内外的成功商人，在享受成功的喜悦之外，向往一种犹如维也纳音乐格调一般的高雅生活，犹如施华洛世奇水晶一般，洗尽铅华，依然铿锵。女主人是全职太太，平时没事的时候会在家里煮煮养生茶，钻研花艺，非常喜欢与女儿交流音乐心得，偶尔也会与男主人一同欣赏从国外挑选回来的艺术品。除此之外，她大部分的时间会与朋友一起逛逛街，做美容，参加各种各样的PARTY，因为女儿经常在家里住，所以一家三口会在阳台上吃点心、谈心、聊艺术……

THE ARISTOCRATS CLASSIC NOBLE MANSION

Jianyu Yongshanjun American Style Country

建宇雍山郡美式乡村

Design Company: Chongqing Pinchen Design
Designers: Pang Yifei, Hu Yajun, Hu Jing
Soft Decoration Designer: Huang Lin
Project Location: Chongqing
Project Area: 380.72 m²
Major Materials: Oman Beige, Fish-belly White Marble, Hangzhou Gray Marble, Black Gold Stone, Archaized Tile, Ash-tree, Teak Wood Floor, Imported Wallpaper, Leather

设计公司：重庆品辰设计
硬装设计师：庞一飞、胡亚君、胡靖
软装设计师：黄琳
项目地点：重庆
项目面积：380.72 ㎡
主要材料：阿曼米黄、鱼肚白、杭灰、黑金花石材、仿古砖、水曲柳、柚木地板、进口壁纸、皮革等

The designer abides by the ideas of providing dreams for a world not having time for dreams, getting people inside the space and constructing an inner world with the institutions of designer. The space used to be a three-dimensional multiple unit, and the designer conceives that the materials applications and space combinations are like musical notes floating in the space. The designer is like a great composer, constructing moving chapters with orders.

The space is presented with euphemistic and restrained approaches, creating some open and generous uncommon temperament, not making people feel cramped at all. The designer relies on the reversibility and continuity of the space structures to acquire the space integrity, trying to change the life modes of inhabitants through the space designs. The design of the whole space appears sedate and luxurious, seeking for the uplifting of inner texture in more sense and appropriately making the aristocratic temperament go through each corner of the space.

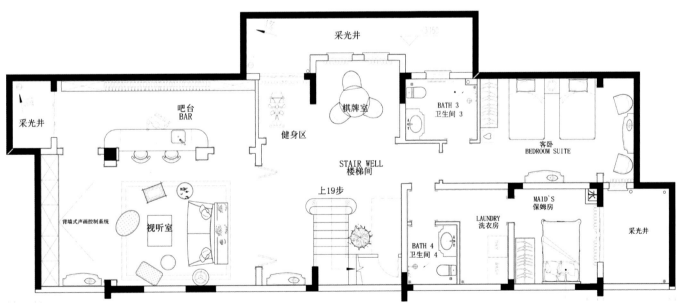

THE ARISTOCRATS CLASSIC NOBLE MANSION

The layout selects bright colors to convey some intensive charms of grand nature, prevailing with American style countryside charms and free and leisurely life style. The pure color tones, the delicate pattern, the arch door and windows, the solid wood furniture and the flowers and grass quietly growing along the sofa, are all singing the free, joyous and casual life in tranquility.

设计师秉承为一个没有时间做梦的世界提供梦想的想法，让自己置身于空间中，利用设计师的直觉来构想一个内心的世界。空间本是一个三维的多面体，设计师构想材料的运用和空间的组合如同音符般漂浮在空间里，设计师如同伟大的作曲家，有序地将其组合成动人的篇章。

空间以一种委婉、含蓄的方式展现出来，给人以开放、宽容的非凡气度，让人丝毫不显局促。设计师借助了空间结构本身的可逆性和连续性去实现整体性，力求通过空间的设计去改变居住者的生活方式。整体空间的设计沉稳、奢华，更多的是追求一种内在质感的提升，恰到好处地把雍容华贵的气质渗透到空间里的每个角落。

在陈设中，选用鲜艳的色彩去传达一种浓烈的大自然韵味，弥漫着美式乡村风情自由休闲的生活方式。纯净的色调、精致的花纹、拱形的门窗、原木家具，在沙发旁静静生长的花草，都在恬静中散发着休闲、喜悦、自由自在的生活情调。

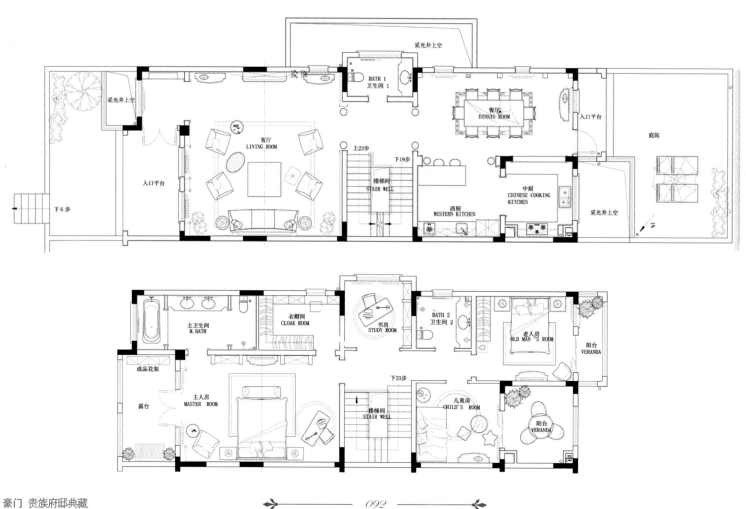

Red Star • Venice Manor Duplex Villa

红星·威尼斯庄园双拼别墅

Design Company: Grand Ghost Canyon Designers Associates Ltd.
Project Location: Changzhou of Jiangsu Province
Project Area: 660 m²
Major Materials: French Beige Marble, Amber Jade, Agate Jade, Afghanistan Gold Marble, Silk Road Beige, Straight Grain Wood Veneer, Rose Gold Steel

设计公司：广州市韦格斯杨设计有限公司
项目地点：江苏常州
项目面积：660 m²
主要材料：法国米黄大理石、琥珀玉、玛瑙玉、阿富汗金、丝路米黄大理石、直纹鸡翅木饰面、丝玫瑰金钢

This is a duplex villa, with traditional French charms as the design orientation for this project. The designer intends to produce a residential space with noble and elegant style. The plane functional planning and space ordering entrust space experiences echoing high-end mansion.

The designer stresses the master-slave relationship in plane graphics, corresponding relationship in space axes and seeks for symmetric and rigid facade treatment. The decorative approaches insist on classical French collocation principles, matching the arch curves of furniture and ornaments, making the space appear elegant and noble. Under the collocations of crystal drop-lights, floor lights and lilies in vase, the romantic and fresh feel comes directly in the face, creating a luxurious and cozy residential space.

The flowers and green plants everywhere, the furniture of careful carving... Everything produces some pastoral

THE ARISTOCRATS CLASSIC NOBLE MANSION

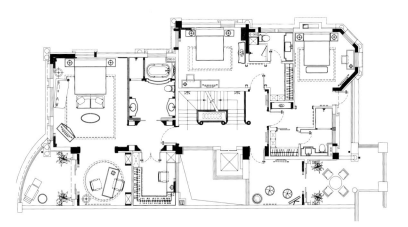
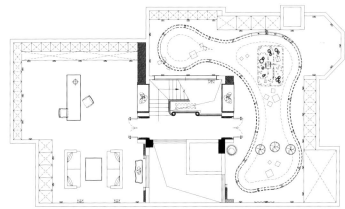

THE ARISTOCRATS CLASSIC NOBLE MANSION

CHIC CITY
Duplex A2-5

中航城复式 A2-5

Design Company: SCD Hongkong Simon Chong Design Consultants Ltd.
Designer: Simon Chong
Project Location: Guiyang of Guizhou Province
Project Area: 206m²

设计公司：SCD（香港）郑树芬设计事务所
设 计 师：郑树芬
项目地点：贵州贵阳
项目面积：206 ㎡

The exquisite lines decorate the traits of youth and the ingenious delicate sculpture gently narrates every aesthetic story touched by each finger, be it elegant, or noble, or pleasing. That is the antique, avant-guard, mix and match leisure life style produced by the designer.

Upon entering the high hall, you can find that the magnificent crystal drop-lights make the space be filled with intoxicating light yellow lights. The carpet full of fashion feel produces rhythmic feel with the combination of light tan, black and white, displaying vividly the artistic conceptions of antique mix and match style, while presenting the integration of eastern and western culture. The high and grand glass window and vertical curtain is like a graceful lady with golden long hair. The sofa has the combination of classical dark gray color and solid wood framework, being pleasing and elegant.

The crystal lights of the living room is noble and elegant, not only echoing the living room's drop lights, but also being in harmony with the dining table in format. The European classical dining chairs seem to be the hero of the whole dining hall, with the cabinet on one side presenting various kinds of foreign wine and red wine. The spacious kitchen specially adds a breakfast table, which can also be used as a bar counter.

In the bright master bedroom space of the second floor, the elegant grand bed is collocated with soft background wall, the subtle and complex pattern carpet adds some profound feel to the space. The nice linen wall bolster softens the hard feel of the whole space, making the bedroom be tranquil and show luxury feel at the same time. In the bathroom, the floor board and wall of marble texture make the solemn and serious feel of the space extend.

The children room closely neighboring the master bedroom applies light tan and dark green, integrating modern and classical elements. The carpet applies graphics full of activity and soft decoration ornaments with European elements, adding some dynamic charms to the rooms.

精致的线条镌染着青春的轨迹，极具匠心的细致雕琢婉约地诉说着每一个指尖碰触过的唯美故事，或高雅尊贵、或雅致静怡。这就是设计师营造出的复古、前卫、混搭的惬意生活方式。

步入挑高的大厅，一盏华丽的水晶吊灯，使空间充满了醉人的淡黄色灯光。富有时尚感的地毯，以驼色、黑色、白色交织出律动感，将复古混搭的意境表现得活灵活现，展现着东西文化的融

THE ARISTOCRATS CLASSIC NOBLE MANSION

合，挑空的玻璃大窗与垂直的窗帘像一位优雅的女子披着金色的长发般美丽。沙发以古典的深灰色与实木框架相结合，舒适而雅致。

餐厅的水晶灯高贵而典雅，不仅与客厅吊灯相迎合，其形状也与餐桌和谐起来。欧式古典的餐椅似乎是整个餐厅的主角，一侧的壁柜展示着各式各样的洋酒、红酒；宽敞的厨房里，还特别增加了早餐台，也可以作为吧台使用。

Jiulongcang Palazzo Pitti Show Flat

九龙仓碧玺样板房

Soft Decoration Furnishing Company: Hangzhou Jishang Design and Decorative Engineering Co., Ltd.
Layout Designer: Ge Xulian
Project Location: Hangzhou of Zhejiang Province
Project Area: 363 m²

软装陈设公司：杭州极尚装饰设计工程有限公司
陈设设计：葛旭莲
项目地点：浙江杭州
项目面积：363 ㎡

The area of this project is 363 m². The whole style is magnificent and luxurious, full of profound dynamic effects. The mastering of the designer towards each detail is quite appropriate. The ingenious color design creates profound elegant atmosphere for the residence. Public spaces such as living room and dining hall focus on warm colors, highlighting the space with blue gray color. The master bedroom continues the warm color of the whole space, with elegant purple gray color as ornaments.

Other than that, the designer boldly selects decorative objects of palace style, with fresh colors and glittering luster. The crystal lights strengthen the visual effects of the space. The whole space reveals the luxurious charms of classical decorative styles, perfectly integrating palace art and contemporary aesthetic elements.

本案面积为363m²，整个风格豪华、富丽，充满强烈的动感。设计师对家的每个细节都拿捏得恰到好处。独具匠心的色彩设计，为居所营造出浓厚的雅致氛围。客厅和餐厅等公共空间以暖色为主，再加入蓝灰色来提亮空间。主卧延续了整个空间的暖色调，加入了高雅的紫灰色作为点缀。

另外，设计师大胆选用宫廷风格的装饰品，色彩娇艳、光泽闪烁，水晶灯则强化了空间的视觉效果。整个空间，透露出古典装饰风格的奢华气韵，将宫廷艺术与当代审美元素完美地融合在一起。

Xinchang Yuelan Mountain Show Flat

新昌悦澜山样板房

Interior Design Company: Hangzhou EHE Interior Design Co., Ltd.	室内设计公司：杭州易和室内设计有限公司
Space Designer: Li Yang	空间设计师：李扬
Soft Decoration Designer: Hangzhou Jishang Design and Decorative Engineering Co., Ltd.	软装设计公司：杭州极尚装饰设计工程有限公司
Layout Designer: Li Shanzhou	陈设设计师：李善洲
Project Location: Shaoxing of Zhejiang Province	项目地点：浙江绍兴
Project Area: 392 m²	项目面积：392 m²

With the familiar design languages, the designer interprets noble and romantic exotic Spanish charms. The Spanish furnishings possess some intensive Mediterranean style, being ardent, free and colorful. Compared with Mediterranean style, the furnishings appear more mysterious, restrained, sedate and profound. The design approaches do not need too many techniques, just maintaining the simple concepts, capturing the lights and getting materials from the grand nature, while boldly and freely making use of colors and formats. The stones get antique treatment, fully representing the true Mediterranean charms, while the furniture furnishings of classical iron art and noble figures decorative paintings on the wall represent the inheritance of pastoral culture. The large open-style study underground is used for storage and presentations functions, while the red wine and cigar display the master's social status and life tastes. The whole space sends out the romance and natural charms of Mediterranean life, while conveying the arrogant temperament of the space from the inner part.

设计师用一贯擅长的设计语言诠释高贵和浪漫的西班牙异国风情。西班牙家居带有强烈的地中海风格，热情洋溢、自由奔放、色彩绚丽。相对于地中海风格，则显得更加神秘内敛、沉稳厚重。在设计手法上，不需要讲究太多的技巧，而是保持简单的信念，捕捉光线，取材于自然，大胆而自由地运用色彩、造型。石材做仿古处理，完全再现真正的地中海风情。古典铁艺类家具的陈设和墙面贵族人物装饰画，体现出庄园文化的传承。地下层的大开间书房，用于收藏展示功能。红酒、雪茄等体现出主人的社会地位和生活品位。整个空间，散发出地中海生活的浪漫与自然情调，并且从骨子里传达出空间的高傲气质。

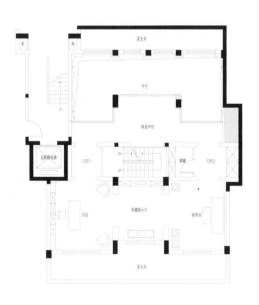
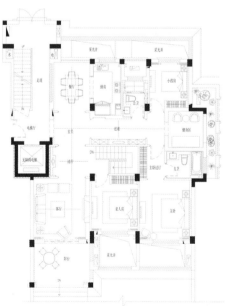
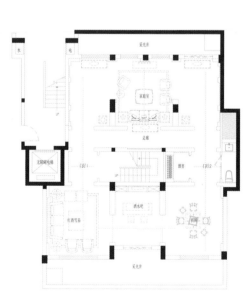

Hanson • spring and Autumn House D+Showflat

恒信·春秋府 D 户型样板房

Design Company: Shenzhen Panshi Max Interiors Co. Ltd., Wu Wenli Design Office
Designer: Wu Wenli、Lu Weiying
Participate Designer: Chen Dongcheng
Project location: Jingzhou
Project Area: 125 m²
Main Materials: Valley sand marble, Blue Moon marble, colorful new grey marble, travertinetile, wall cloth, leather, Rome, South America America teak, walnut wood floors, merbau outdoor wood, gray glass, mirror, bright black ash

设计公司：深圳市盘石室内设计有限公司吴文粒设计事务所
设 计 师：吴文粒 陆伟英
参与设计：陈东成
项目地点：荆州
项目面积：125 m²
主要材料：蓝金砂大理石、新月亮谷大理石、多彩灰大理石、罗马洞石瓷砖、墙布、皮革、南美柚木、美国胡桃木实木地板、菠萝格户外木、灰玻、灰镜、亮面黑钢

The highest realm of Oriental art style, the pursuit of write and draw freely as one wishes, a synthetic and vigorous and simple artistic attainments. In this case, the designer intended to achieve the same level, communication space atmosphere, simple and elegant atmosphere. The traditional Eastern spirit contained in the interior space, through China traditional architectural symmetry plane layout and facade composition, the entire space using the traditional Chinese red, continuation of traditional Chinese style screen, and original furniture design, showing the modern Chinese space atmosphere.

Indoor everywhere decorated with Designer displays, such as furniture, lighting and accessories, all kinds of exquisite display China objects, these objects and modern design in the space encounter appropriate, make modern design contains deep China historical cultural imprinting, let old Chinese artistic vitality, grasp, use and composition the designer of the Chinese elements accurate. An oriental cultural background, has now, fashion model space. In conveying the distinguished atmosphere at the same time, let a person to feel Oriental cultural relics. Subversion of the traditional culture in developing at the same time, the deep, inject new

elements, revealed the Chinese Zen and the simple Wenxin elegant space, each scene and angle space makes people feel the joy. The visual, another kind of enjoyment, to achieve "a few people outside, sit listening to the spring bird" realm.

东方艺术的最高境界是写意，追求挥洒自如、一气呵成和苍劲质朴的艺术造诣。在本案例中，设计师意图达到同样的境界，传达大气、质朴及优雅的空间氛围。把东方传统的气韵蕴藏在室内空间中，整个空间运用传统中国红，延续传统中式屏风，以及精心设计的原创家具，呈现出现代中式空间的大气。

室内处处都摆放着设计师精心设计的陈列品，如家具、灯饰和挂件，陈列着各种精致的中国物件，这些物件与现代的设计在恰当的空间邂逅，让现代设计蕴涵着深深的中国历史文化印迹，让古老的中国艺术焕发活力，设计师对中国元素准确的把握、运用和组合。呈现出一个具有东方文化底蕴，兼具现代、时尚的样板空间。在传达尊贵大气的同时，让人切实地感受东方的人文古韵。在弘扬中华深厚文化底蕴的同时，颠覆传统，注入新的元素，温馨优雅的空间中透露出中式的禅意和朴实，空间中每个景致与角度都使人感觉到喜悦。使人视觉上得到另外一种享受，达到"寥寥人境外，闲坐听春禽"的境界。

THE ARISTOCRATS CLASSIC NOBLE MANSION

Villas du Lac, Building B, Show Flat No. 2

天悦湾 B 栋 2 样板房

Design Company: STEVE.S DESIGN
Designer: Shi Lirui
Project Location: Shenzhen
Major Materials: Marble, Titanium, Piano Lacquer, Gold Mirror, Leather, Wallpaper

设计公司：史礼瑞设计师有限公司
设 计 师：史礼瑞
项目地点：深圳
主要材料：云石、钛金、钢琴漆、金镜、扣皮、壁纸

This space makes use of the clashing of different colors, making abundant colors leave people with refreshing visual impressions. Although the host is a businessman, yet he is quite traditional, and quite stresses family. The ingenious integration of host's likes and European classical styles make the host get free breath here.

Entering the living room is like walking into the tranquil oriental country, yet with modern style and elegance. The designer applies black and dark coffee color as the tone for the whole space, making the space produce intensive visual impressions through the contrast of different materials and mirror surfaces. The oriental elements are collocated with flamboyant crystal accessories, representing the vigor of modern atmosphere.

In the bedroom, the layered beige color tone, silk bedding accessories and appropriate lighting finely set off the temperament of classical furnishings, making the space acquire warm charms.

THE ARISTOCRATS CLASSIC NOBLE MANSION

这个空间运用不同颜色的碰撞，让斑斓的色彩给人耳目一新的视觉感受，男主人虽然是个生意人，但比较传统，而且非常注重家庭。主人的喜好与欧式新古典巧妙的交融，让居室主人可以在这里得到自由的呼吸。

走进客厅，仿佛走进了悠远宁静的东方国度，却又不失现代的摩登与优雅。设计师运用黑与深咖色作为整体空间的基调，通过不同材质和镜面的对比，让空间产生强烈的视觉映像。东方元素配以流光异彩的水晶器皿，体现了现代气息的活力。

在卧室中，叠加的米褐色调、丝质的床品、恰到好处的灯光，极好地烘托出经典家居的气质，令这里充满温馨的情调。

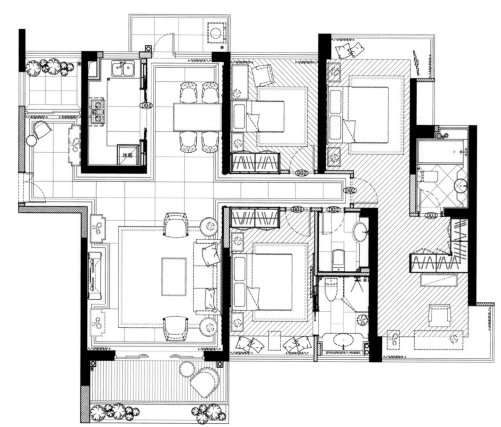

THE ARISTOCRATS CLASSIC NOBLE MANSION

THE ARISTOCRATS CLASSIC NOBLE MANSION

Central Building Garden Villa

央筑花园洋房

Design Team: KSL Design FirmChief
Space Designer: Lin Guancheng
Soft decoration design company: Shenzhen Yisan Yisan Decoration Co., Ltd.
Soft decoration designer: Wen Xuwu
Project Location: Huizhou of Guangdong Provice
Project Area: 210 m²
Main Materials: Oak, bronze, artificial stone, ebony floor
Photographers: Jianghe Photography

设计团队：KSL 设计事务所
空间设计师：林冠成
软装设计公司：深圳壹叁壹叁装饰有限公司
软装设计师：温旭武
项目地点：广东惠州
项目面积：210 ㎡
主要材料：橡木、古铜、人造石、黑檀木地板
摄　　影：江河摄影

The Chinese style of the ancient charm and elegant tolerance always makes people fascinated, but modern urban people hard to handle the old and dull in traditional Chinese style. The case adopted rich presents classical Chinese elements and ameliorated to form the modern Chinese new style. Massive oak materials used to create the natural and comfortable style and whole space was designed skillful tact and a degree of relaxation.

Living room: There are a lot of natural wood materials in the living room design materials and colors and patterns of wood itself to create a wonderful home atmosphere. Ebony floor deep color matched with the bright tone of ceiling makes the whole of space nature and harmony. From the spatial layout to the furniture modeling, hale line have formed a regular space, decorative painting of the wall enhance the artistic conception of space.

Bedroom: Space atmosphere of bedroom manifests more warm and comfortable. Wood veneer walls makes people the peace of mind, horizontal or vertical geometric lines appeared from time to time in the ceiling, walls and furniture, through time and space of the ancient feeling emerged, comfortable and warm.

Stairs: Staircase design most abstain from dim. Light wood floors and bright light and lamps instantly brightens the whole space. The whole wooden stripe decorating the walls formed a perfect partition, at the same time, brought the elegant natural breath, matched under the light looks like bright and fresh.

Dining room: Concise form table was chosen dark brown color and a bouquet of blue flowers decoration created romantic space atmosphere.

Kids room: Dark green and stripes are the active elements of kids room. From the choice of curtains color to bed goods are conscientious and meticulous. There are fresh with playful and peace with warm.

中式风格的古朴韵味和典雅气度总令人心神往之，但传统中式的古旧与沉闷却常令现代都市人难以驾驭，本案采撷了丰富的经典中式元素，将之改良形成全新的现代中式风格。大量橡木材质的使用营造出自然舒展的格调，整体空间设计手法纯熟老练，张弛有度。

客厅：天然木材大量地出现在客厅的设计材料中，木料本身的色彩和纹饰营造了温良敦厚的居家氛围。黑檀木地板的深沉色彩配以顶棚的明亮色调，整体自然而和谐。从空间布局到家具造型，硬朗的线条都形成了规整的空间感，墙壁的装饰画则提升了空间意境。

卧室：整体色调一致的基础上，卧室的空间氛围更突显出温馨与舒适感，木纹贴面的墙壁让人心神安宁，或横或竖的几何线条不时出现在顶棚、墙壁与家具上，穿越时空的古韵感扑面而来，舒展而温暖。

楼梯：楼梯设计最忌暗淡，浅色的地板与璀璨的灯饰瞬间提亮了整个空间，通体木制条纹装饰的墙壁则在形成完美分区的同时，带来淡雅的自然气息，在灯光的配合下显得明亮而清新。

餐厅：简洁造型的餐桌选用了深咖色系，一束蓝色的鲜花装点出浪漫的空间氛围。

儿童房：墨绿与条纹是儿童房的活泼元素，从窗帘色调到床品选择均一丝不苟，清新中带着俏皮，宁静中带着温馨。

THE ARISTOCRATS CLASSIC NOBLE MANSION

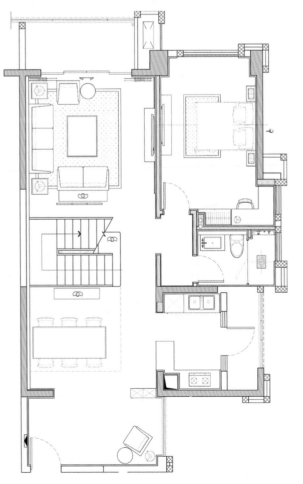

THE ARISTOCRATS CLASSIC NOBLE MANSION

Antique Yuppie · Longevity Show Flat

复古雅痞士·长寿样板间

Design Company: Chongqing Pinchen Decorative Design and Engineering Co., Ltd.
Hard Decoration Designers: Pang Yifei, Li Jian, Liu Liping
Soft Decoration Designer: Cheng Jing
Project Location: Chongqing
Major Materials: Stone, Hard Roll, Mirror Surface, Stainless Steel, Wood Floorboard, High Luster Board
Project Area: 330 m²

设计公司：重庆品辰装饰设计工程有限公司
硬装设计师：庞一飞、李健、刘丽萍
软装设计师：程静
项目地点：重庆
主要材料：石材、硬包、镜面、不锈钢、木地板、成品亮光板等
项目面积：330 ㎡

The intensive antique charms of this project have peculiar aesthetic feels, and the Yuppie attitudes go deep into every corner of the space. The mottled and time feel is just like the artistic conceptions created in the movie The Legend of 1900. The designer makes use of artistic presentation approaches to convey this time feel, making people feel the beauty and comfort.

The space master seeks for fashion life, has avant-guard thinking, knows about enjoyment and is passionate towards emotions. For him, ideal residence should be like this and study, studio and billiard parlor are all necessary. Urban map, magnifying glass of detective, electric guitar and projector are all there. The DIY keys hanging painting on the wall is the secret to start happy life, and the antique clock on the wall is always reminding us to cherish the warm time of family. Each object has its own story, awaiting you to write.

Home is everyone's dream space. As a space, it has its own moods and characteristics. Home is a constantly regenerating

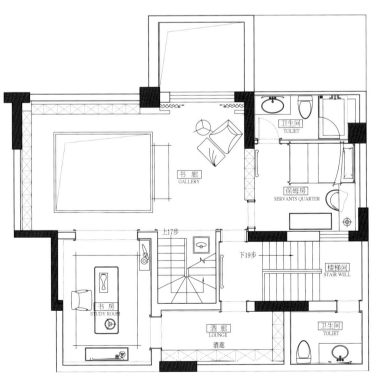
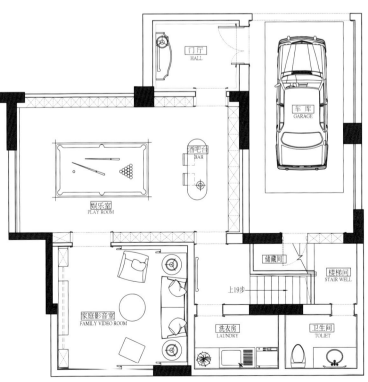

THE ARISTOCRATS CLASSIC NOBLE MANSION

THE ARISTOCRATS CLASSIC NOBLE MANSION

dream. It tells us that, we should not rush to make it change into anything, we just need to follow our hearts. What it deserves would come around.

本案浓郁的复古情调有着独特的审美感，雅痞的态度深入空间的每一个角落。所营造的斑驳感、时代感仿佛电影《海上钢琴师》所营造的情怀。设计师用艺术的表现手法来传达这种时代感，让人感到美，感到舒适。

这样空间的主人追求时尚生活、思想前卫、懂得享受，对待感情一往情深。在他们心目中，理想居所应该是这样：书室、画室、台球室每一间都必不可少。城市地图、侦探的放大镜、电子吉他、放映机，一样也不能少。墙上挂着的DIY钥匙挂画是开起幸福人生的秘密，书桌上的复古时钟始终提醒人们要珍惜家庭中的温馨时光。每一件装置都有他的故事，等待你去谱写。

家是每个人的梦想空间，它是空间，也有心情和性格。家是一个不断被更新的梦想，它告诉大家，不要急于让它变成什么样子，只要顺着自己的心意走，该有的总会有。

THE ARISTOCRATS CLASSIC NOBLE MANSION

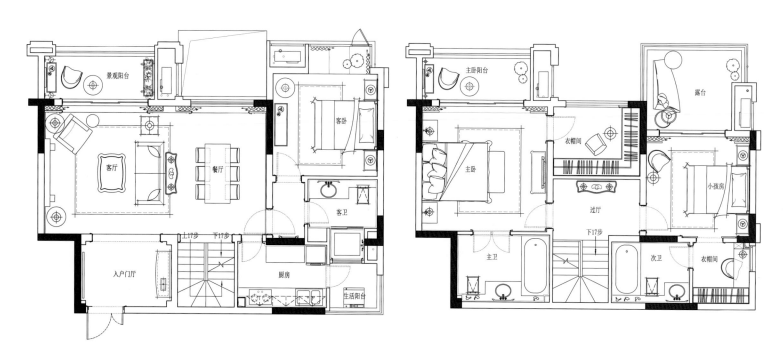

豪门 贵族府邸典藏

THE ARISTOCRATS CLASSIC NOBLE MANSION

Hongji Ziyuan J1 Show Flat

鸿基紫韵 J1 示范单位

Design Company: PINKI Interior Design & IARI Interior Design Co., Ltd.
Designer: Liu Weijun
Project Location: Xi'an of Shanxi Province
Project Area: 230 m²
Major Materials: US White Oak, Jazz White Marble, Iceland Gray Marble, Hand-painted Wallpaper, Silver Foil

设计公司：PINKI 品伊创意集团 & 美国 IARI 刘卫军设计师事务所
设 计 师：刘卫军
项目地点：陕西西安
项目面积：230 ㎡
主要材料：美国白橡木、爵士白大理石、冰岛灰大理石、手绘壁纸、银箔

"Shi"

Shi in Chinese can be understood as release or interpretation, releasing some elegant, brisk and fashionable artistic atmosphere, interpreting the contrast of black and white and clash of passions, presenting the high-taste life concept of the property owner.

The host of the project is a composer and the hostess is a Ballet Dancer, who both like a living environment which is fresh, bright, comfortable and full of cultural connotations. In order to present the host's requirements on life quality, and to excavate the space's business values and practical quality, the designer changes the original garden balcony into a dining hall, the 270°outdoor views are brought to you, quite pleasing.

Other than that, the study of master bedroom is extended to

the balcony, producing a multi-functional room, where one can play chess, drink tea and talk with others, which not only bring some fun for the host's life, but also strengthen the space feel of the study.

In the creation of space atmosphere, the furnishings play an important role. Wallpaper of black and white pattern, concise white fabric sofa, white leather tea table, white fashionable chairs, collocated with crystal drop-lights and white glass table lamps of black pattern, construct the modern and

fashionable artistic atmosphere for the living room. The glass candlesticks, white porcelain tea cups and tea pots and metal plate on the tea table inside the living room light up the space. The master bedroom has ballet skirt, ballet figure sculpture and soft sofa of blue and white pattern, all are residing beside the concise floor lamp. In the study, the fashionable suitcase and painting, classical phonograph and other ornaments, all are interpreting the elegant atmosphere of this art family!

The Cairn Hill, Top Level Duplex Mansion

朗月峰顶层复式宅

Design Company: Simon Chong Design Consultants Ltd.
Designer: Simon Chong
Project Location: Jiulong West Zone, 2Hong Kong
Project Area: 530 m²

设计公司：SCD(香港)郑树芬设计事务所
设 计 师：郑树芬
项目地点：香港九龙
项目面积：530 ㎡

This top level duplex mansion has great geological location. Due to the location being close to west peak, you can have a full view of outside landscape, as well as mountain views and ocean views. Together with some islands, it is quite a magnificent view here.

The top level of the project makes use of the space height advantage, with European classical lines as the major decorations. The whole space has pure white color as the tone, being grand and bright visually, and demonstrating elegant and noble temperament. Part of the wall applies elegant European wallpaper as the decorations, allowing each space to have variations and creating refreshing sensations for people.

The highlight of the design is also presented on selection of furniture, the designer applies some furniture brands used in French Presidential Palace – Hugues Chevalier and British brand Christopher Guy. Hugues Chevalier(HC) is from the family of Louis Vuitton, which has the features of Art Deco and post-modernism, together with sensible and peculiar characteristics, as a furniture brand winning love of rich and powerful people and aristocrats from Europe and America. Bill Gates, Beckham, Schumacher and boxing champion Ali

all have this brand of furniture at home. As for soft decoration design, the designer selects French top brand chandeliers and furnaces to make each space be brilliant and shining.

Another advantage of top level duplex mansion is the wide balcony, which is set with 18m-wide vertical garden as the key decoration. Different outdoor furniture also have divisions towards the whole balcony, presenting different functional atmosphere, where the property owner can invite friends to enjoy the fun of outdoor party.

此套顶层复式宅邸的地理位置极佳，由于靠近西部山顶，室外的景观一览无遗，同时可以俯瞰山景和海景,加上部分港岛，形成了非常壮丽的景色。

项目的顶层利用了楼顶高挑的优势，将欧式古典的花线作为主要的装饰。整个空间以纯白色为基调，在视觉上宽敞而明亮，彰显高雅大方的气质。部分墙面利用典雅的欧式壁纸进行点缀，使每一个空间都有变化，给人焕然一新的感觉。

设计的亮点还体现在家具的选用上，设计师采用了法国总统府使用的家具品牌——Hugues Chevalier，以及英国品牌 Christopher Guy。Hugues Chevalier(HC)来自路易威登家族，它以Art Deco和后现代风格为特点，兼具感性以及独特的个性，是欧美富豪、贵族热衷享用的家具品牌。比尔·盖茨、贝克汉姆、舒马赫、拳王阿里加都使用该品牌的家具。在软装设计上，设计师选用了法国顶级品牌的吊灯和壁炉，使每一个空间都璀璨而闪亮。

顶层复式住宅的另外一个优点是天台。宽敞的天台设置了18m宽的垂直花园作为重点来装饰，不同的户外家具也将整个天台进行区分，展现出不同的功能氛围，业主可尽情邀请朋友在这享受户外party的乐趣。

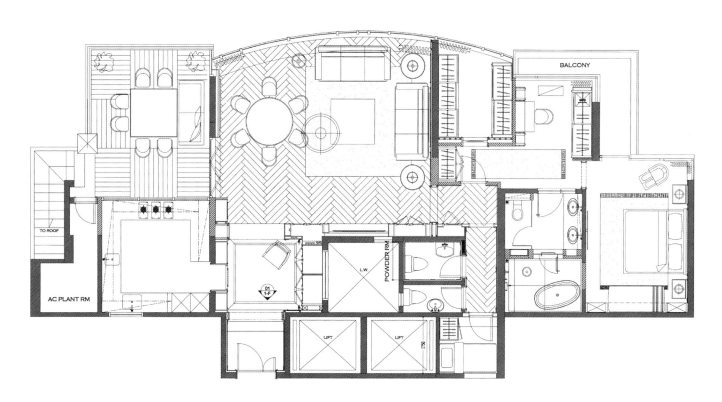

THE ARISTOCRATS CLASSIC NOBLE MANSION

THE ARISTOCRATS CLASSIC NOBLE MANSION

Hongji Ziyun J2 Show Flat

鸿基紫韵 J2 示范单位

Design Company: PINKI Interior Design & IARI Interior Design Co., Ltd.
Designer: Liu Weijun
Project Location: Xi'an of Shanxi Province
Project Area: 220 m²

设计公司：PINKI 品伊创意集团 & 美国 IARI 刘卫军设计师事务所
设 计 师：刘卫军
项目地点：陕西西安
项目面积：220 ㎡

Hiding

Xi'an used to be called Chang'an in the ancient time, as one of the four ancient capitals of civilization, with a long history and abundant cultural sediments. "The old capital with Qin Dynasty charms, and get satisfied in the ancient town." Dragon in China, root in Xi'an, that is the perfect interpretations for Xi'an!

This project has "hiding" as the theme and produces brisk contrast between "hiding" and "releasing". The design has blue, red and yellow as the tone colors, representing the space's solemn and profound quality. The host of this project is an architect, who enjoys collecting red wine. The hostess is a media supervisor, who always show up in some fashion parties, who also likes dancing and music, thus the whole space is full of artistic atmosphere.

THE ARISTOCRATS CLASSIC NOBLE MANSION

The structure of this project is quite square, with the original bar counter changed into a wine collection area, meeting with the host's likes for red wine. There is a multi-functional room between the bedroom and the balcony of the living room, where one can entertain and meet friends here. The dark color wood veneer surrounds the whole space, presenting low-profile and luxurious feel. The crystal drop-lights of black mask shyly shine on the blue soft leather sofa, which seems like a love story is put on the stage. The blossoming red crape myrtle flowers in the transparent glass vase have some influence on the surrounding air and white coral reef.

The master bedroom's blue stripe wallpaper and the red leather bedside echo the tone color of the living room. There are dressing mirror, evening dress, hat, handbag and glass slipper in the corner, thus it is like the hostess is just back from a party. The unique cultural and artistic infections are presenting the spectacular romance exclusive to the time.

隐

西安，古称长安，是世界四大文明古都之一，具有悠久的历史和浓郁的文化积淀。"秦风韵故都，满意在古城"，龙在中国，根在西安，是对西安的完美诠释！

本案以"隐"为主题，将"隐"与"释"形成鲜明的对比。主要以蓝、红、黄为主色调，体现空间庄重、沉稳的品质。本案的男主人为建筑师，喜欢收藏红酒，女主人为媒介总监，经常出席一些时尚Party，同时喜欢舞蹈和音乐，因此整个空间处处流淌着艺术气息。

本案的结构比较方正，将原有的吧台变为藏酒区，满足男主人对红酒的钟爱。卧室与客厅的阳台之间搭建出一个多功能房，可以在此娱乐及会友。深色调的木饰面包围着整个空间，彰显着低调与奢华感。黑色面罩的水晶吊灯羞答答地照射着蓝色的软皮沙发，好似在上演着一段爱情故事。装在透明的玻璃瓶内盛开的红色紫薇绽放着、感染着周边的空气和白色的珊瑚礁。

主卧的蓝色条纹壁纸、红色皮质床头与客厅的主色调相呼应，角落的试衣镜和晚礼服、帽子、手袋、水晶鞋好似女主人刚参加完一个Party。独到的文化艺术感染力，自然流露出岁月中那份特有的浪漫。

THE ARISTOCRATS CLASSIC NOBLE MANSION

Villas du Lac, Building B, Show Flat No. 1

天悦湾 B 栋 1 样板房

Design Company: STEVE.S DESIGN
Designer: Shi Lirui
Project Location: Shenzhen
Major Materials: Lacquer, Marble, Woolen Blanket, Hand-painted Wallpaper, Fabrics, Wood Floor

设计公司：史礼瑞设计师有限公司
设 计 师：史礼瑞
项目地点：深圳
主要材料：手扫漆、大理石、羊毛毯、手绘壁纸、布艺、木拼地板

As the host and hostess both have overseas life experiences, they both long for free and romantic life artistic conceptions and atmosphere. After they get back homeland and settle down, the residences are used for fostering heart and soul. The space can not be that big, yet everything inside should all be delicate, creating holiday-like pleasure.

In hard decorations, the space is simplified European tones. The combination of white, beige and straight lines and grids is bright and brisk. For latter-stage furnishings, the bouncing colors and delicate little ornaments make the whole residence avoid simple monotonousness, and get rid of complexity. From the space layout, we can see that the property owners retain the western life style. The open-style moving lines make their interactions become more intimate. Without separations of objects, the designer makes use of furniture, plants or lights to produce end views.

The refined materials make the space become delicate and aesthetic. The collocations of light and dark colors make the space perform elegant atmosphere, without much too complicated graphics, or abundant colors, or luxurious expressions, this project creates inexpressible quality feel and comfort feel.

由于男女主人都有在国外生活的经验，非常向往自由浪漫的生活意境和氛围。回国定居，居室就是用来修养身心的，空间可以不是很大，但是里面的一切都应该是精致的，可以带给他们度假般的惬意。

空间在硬装上，是简化过的欧式格调，白色、米色以及直线条和方格的组合，明朗而利落。后期陈设上，加入跳跃的色彩及精致的小饰品，让这个居所在整体上既避免了简单的无味，又抛弃了繁琐的累赘。空间布局上可以看出男女主人还保留着西方的生活方式，开放式的动线可以让他们的互动更为密切。没有实体的区隔，而是运用家具、植物或者光线制造端景。

讲究的材质将空间衬托得精致唯美，深浅色调的搭配营造出空间的优雅气息，没有繁复的图案，没有丰富的色彩，也没有奢华的表达，却给人一种无以言表的品质感与舒适感。

THE ARISTOCRATS CLASSIC NOBLE MANSION

THE ARISTOCRATS CLASSIC NOBLE MANSION

Catic City B3-2

中航城 B3-2

Design Company: SCD Hongkong Simon Chong Design Consultants Ltd.
Designer: Simon Chong
Layout Designers: Simon Chong, Du Heng, Yang Li
Project Location: Guiyang of Guizhou Province
Project Area: 103 m²

设计公司：SCD（香港）郑树芬设计事务所
设 计 师：郑树芬
陈设设计师：郑树芬、杜恒、杨立
项目地点：贵州贵阳
项目面积：103 ㎡

The elegant artistic elements integrate with peculiar aesthetic feels, the designer makes the space over 100 square meters become a delicate and romantic life sphere. The whole project is full of various kinds of French style artistic elements. And the application of various kinds of accessories and dining wares produces a warm and restrained urban life space. In the fascinating dining hall, the solid wood dining table and chairs of European antique style echo each other, and send out some refreshing and elegant atmosphere with the ornament of light green colors. The designer connects the guest room and dining hall as a whole. The TV background wall of the living room applies artistic glass veneers, presenting some spacious and cozy feels. The exquisite crystal drop lights and large area of light color wood veneer wall combine with French lines, making the whole space appear softer.

THE ARISTOCRATS CLASSIC NOBLE MANSION

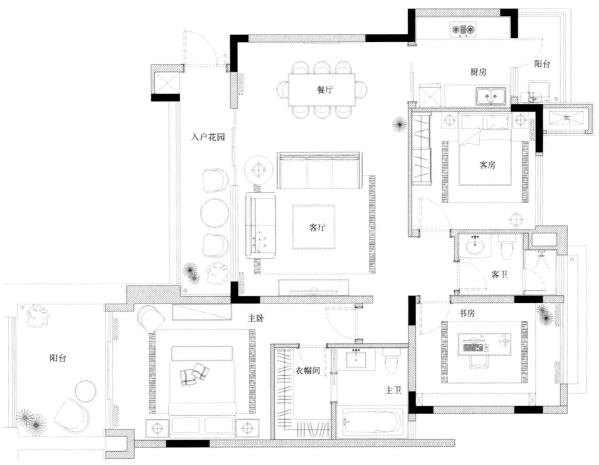

THE ARISTOCRATS CLASSIC NOBLE MANSION

清雅的艺术元素与独特的美感相融合,设计师将一百多平方米的空间演绎出了一种精致而浪漫的生活意境。整个项目充满了各种法式的艺术元素,各种摆件及餐具的运用,营造出了一个温馨而内敛的都市生活空间。引人注目的餐厅里,实木餐桌和欧式复古风格的椅子相互衬托,加上淡绿的点缀,透露出清新淡雅的氛围。设计师将客厅、餐厅连成一线,客厅的电视背景墙采用了艺术玻璃饰面,呈现出宽阔舒适的感觉,精美的水晶吊灯、大面积的浅色木饰墙面结合法式的线条,使整个空间更加柔美。

Zhongzhou Central Park, Phase 2, 9-B01

中洲中央公园二期 9-B01

Design Team: KSL Design Studio
Leading Designer: Lin Guancheng
Project Location: Shenzhen
Project Area: 238 m²
Major Materials: Ash-tree, Bronze, Jazz White Marble, Wall Cloth, Leather

设计公司：KSL 设计事务所
主设计师：林冠成
项目地点：深圳
项目面积：238 m²
主要材料：水曲柳、古铜、雅士白、墙布、皮革

The romantic charms of France leave people with limitless imaginations, with laudable soft curves and delicate and refined handwork. For this project, the horizontal and vertical lines and orderly space divisions make the space full of rational emotions, the colors of bronze and jazz white make people have quiet moods and leisurely and sedate atmosphere pervade inside the space. The application of brown velvet and leather highlights the noble magnificence and grandeur.

Living Room: The decorative paintings of pure lines are full of modern temperament, the color tone of the whole living room is elegant and not showy, the tone colors of gold brown and bronze are restrained and sedate. The same color lighting accessories and sofa of velvet materials match carpet of geometric knitting, making the space appear elegant and delicate.

Dining Hall: with white as the bottom color, the decorative colors of gold and bronze highlight the magnificent temperament of spaces and the antique mid-century decorative elements go throughout the furniture and dining accessories of European style, and the whole dining environment appears elegant and graceful.

Study: the design of study gets inspirations from European classical art, the selection of furniture is unified with the whole of dining hall and the collocations of jazz white marble and leather make the whole space full of North-European charms.

Bedroom: the geometric patterns randomly spread inside the details of the space, which can be sliding door, or TV background wall. The warm and stretching color tones and the silk bedding accessories of gold coffee color bring out the best in each other. Without too much decorations, the gentle and cozy atmosphere naturally spread inside the space.

Child's Room: the check wall of light coffee color is collocated with decorative painting of candy color does not make the child's room be lofty inside the whole design, and finely coordinates the whole style, being dynamic and low-profile.

法兰西的浪漫情怀让人充满想象，柔美的曲线和精雕细琢的做工值得称赞，在本案中，横平竖直的线条和规整的区间划分又让空间充满了理性的情感，古铜、雅白相间的色彩让人心情平静，悠然而稳重的气息也弥漫其中。棕色丝绒和皮革的运用凸显了贵族的大气与华丽。

客厅：纯线条的装饰画颇有现代的气质，整个客厅的色调典雅而不张扬，金棕色与古铜为主打的色调内敛而沉稳。同色系的灯饰与丝绒面料的沙发配合几何针织的地毯，令空间典雅而不失精致感。

餐厅：白色打底，金色与古铜的装饰色彩提升了空间的华丽气质，复古的中世纪装饰元素弥漫于欧式风格的家具与餐具之中，整个就餐环境雅致而优美。

书房：书房的设计从欧洲古典艺术中吸取灵感，家具的选择与餐厅整体一致，雅士白与皮革的搭配令空间颇具北欧风情。

卧室：几何图案不时出现在空间的细节之中，既可以是推拉木门也可以是电视背景墙，温暖而舒展的色调与金咖色的丝绸床品相得益彰。无需过多的装饰，温软舒适的气息自然弥漫开来。

儿童房：浅咖色的格纹墙壁配以糖果色的装饰画，既不让儿童房突兀于整套设计，又很好地协调了整体风格，生动而不张扬。

Poly Zhonghui Garden, Building 8, House Type D

◆◆ 保利中汇花园8座D户型 ◆◆

Design Company: Guangzhou Daosheng Design Co., Ltd.	设计公司：广州道胜设计有限公司
Designer: Tony Ho	设 计 师：何永明
Associate Designer: Daosheng Design Team	参与设计：道胜设计团队
Project Location: Foshan of Guangdong Province	项目地点：广东佛山
Project Area: 160 m²	项目面积：160 ㎡
Major Materials: Gray Marble, Gray Onyx, Jazz White Marble, Black Ebony Composite Floorboard, Straight Grain Wood Veneer, Champagne Gold Stainless Steel, Wallpaper, Leather	主要材料：灰洞大理石、清水玉大理石、雅士白大理石、黑檀复合木地板、科技酸枝直纹木饰面、香槟金不锈钢、壁纸、扪皮
Photographer: Peng Yuxian	摄 影 师：彭宇宪

This project makes use of fashionable and magnificent presentation approaches to create modern style design for the space with international visions. The mutual reflection of stainless steel and dark gray paint appears sedate and unique, making the picture display luxury in restrained quality, and elegance in sedateness. This project interprets the fashion and luxury of city, while producing warm furnishing atmosphere.

The whole space is dynamic and transparent, with dark color wood veneer and light gray stones producing an international fashionable residential space. The soft decorations are collocated with warm and magnificent orange color to highlight the visions, and dark gray stoving varnish, furniture of stainless steel and hard stainless steel closing echo each other, making the whole space be harmonious and integrated. The decoration cabinet of the corridor breaks the narrow, long and dark shortcomings of corridor, allowing the space to acquire rhythmic feel.

Inside the dining hall, the luxurious crystal droplights are some bright scenery lines, with magnificent lights shining

THE ARISTOCRATS CLASSIC NOBLE MANSION

Longquan Mansion, Building 2, House Type 2102

龙泉豪苑 2 栋 2102 户型

Design Company: Grand Ghost Canyon Designers Associates Ltd.
Project Location: Dongguan of Guangdong Province
Project Area: 560 m²
Major Materials: Babylon Stone, Saint Laurent Black Gold Stone, Mosaic Tile, Solid Wood Floor Board, Camphor Tree Wood Veneer, Rose Gold Marble, Gray Glass, Carpet, Imported Wallpaper, Leather

设计公司： 广州市韦格斯杨设计有限公司
项目地点： 广东东莞
项目面积： 560 m²
主要材料： 巴比伦石材、圣罗兰黑金石材、黄碟贝马赛克、实木木地板、香樟树榴木饰面、玫瑰金、灰玻璃、地毯、进口壁纸 、皮革等

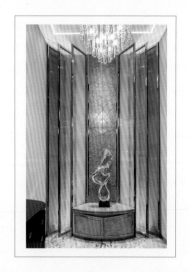

This project is located in Dongguan Humen Town which is No. 1 of China's top 100 towns. As the core area for the future development of Humen, the location of this project has quite complete surrounding life supporting facilities, with education, medical service, transportation and grand business shopping square at the door.

This 560 m² flat has the design with two halls, seven bedrooms and three balconies, with house type fully considering the integrity of functions and format. The designer applies the symmetric design approaches, with the design element of "W" going throughout the space. A large area of the ground applies Babylon stones, with 45°inclined pattern treatment, and the ceiling format also applies the same pattern treatment. The cloakroom combines the requirements for daily life, with adjustable hanger, quite convenient for daily use of the host. The ceiling combines the advantages of 3.8 m high, adjusting the air-conditioner's air

THE ARISTOCRATS CLASSIC NOBLE MANSION

THE ARISTOCRATS CLASSIC NOBLE MANSION

THE ARISTOCRATS CLASSIC NOBLE MANSION

Concise Gentility · Catic City B2-2

简约绅士主义·中航城 B2-2

Design Company: SCD Hongkong Simon Chong Design Consultants Ltd.
Designer: Simon Chong
Project Location: Guiyang of Guizhou Province

设计公司：SCD（香港）郑树芬设计事务所
设 计 师：郑树芬
项目地点：贵州贵阳

With the gentle temperament,
Concise, hale and noble,
Delicate and meticulous,
What we perceive,
Is the sedate and restrained spirits...

Elegant temperament and high-end life quality is the motif of this project. Living inside this space, you can feel more leisure and romance. The whole space attains intensive gentleman temperament. The designer makes use of elegant, or strong colors, or exquisite format to produce refreshing decorative effects. The whole life space send out some leisurely comfort quality, and with collocations of popular design elements. The ingenious soft decorations are quite unique.

THE ARISTOCRATS CLASSIC NOBLE MANSION

一种彬彬有礼的气质，
简练、硬朗、高贵，
精致到一丝不苟，
我们感受到的，
就是沉稳内敛的精神……

优雅的气质与高端的生活品质感是本案的表现主旨。生活在这个空间中，感受到更多的是惬意与浪漫，整体空间具有强烈的绅士气息。设计师用或雅致、或浓烈的色彩、精美的造型来塑造令人耳目一新的装饰效果。整个生活空间彰显出一种安逸的舒适性，又展现出流行设计元素的搭配，巧妙的软装设计令人耳目一新。

Xiamen Poly Ocean May Flower

厦门保利海上五月花

Design Company: Sunny Partnership • Space Planning
Designers: Sunny Wang , Robin Wang
Project Location: Xiamen of Fujian Province
Project Area: 380 m²
Major Materials: Wood Veneer, Leather, Wallpaper, Stone

设计公司：尚诺柏纳·空间策划
设 计 师：王赟、王小锋
项目地点：福建厦门
项目面积：380 ㎡
主要材料：木饰面、皮革、壁纸、石材

The design theme of this project is "Sunshine Cryptolalia of Florida," with as design background Florida which enjoys the name of "Sunshine State." For this project, the designer aims to fully release romantic atmosphere of sunshine and grayish blue coast, producing simple and joyful atmosphere brought by slow rhythm. The materials emphasize on environmental quality. The manual work stresses on delicacy. The whole atmosphere focuses on warmth, comfort and leisure of family, trying the best to produce some quality life, and manifesting the taste of the property owner.

Stepping into the inner room, one feels pleasant to see the furniture in black, white, gray and brown. The design releases a feeling of calm and elegant. The whole design emphasizes on decorative effect. The designer uses modern way to show classic mould, the sunshine and romantics has been fully revealed.

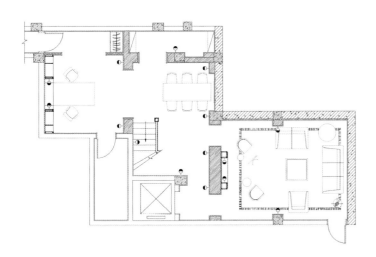
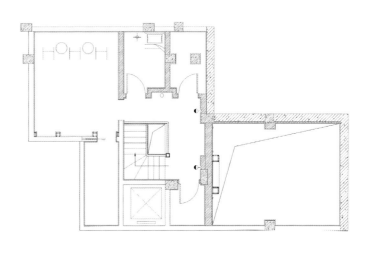

THE ARISTOCRATS CLASSIC NOBLE MANSION

THE ARISTOCRATS CLASSIC NOBLE MANSION

本案的设计主题是"佛罗里达州阳光密语",以有"阳光之州"之称的佛罗里达作为设计背景。在本案中,设计师旨在充分强调阳光、灰蓝海岸的浪漫气息,营造慢节奏生活所带来的简单、快乐的氛围。用材强调环保,做工强调精致,整体氛围着重表现家的温暖、舒适和休闲,极力打造一种赋于格调的生活,彰显业主的独特品位。

步入室内,会让人觉得一阵莫名的惊喜,黑、白、灰、棕的主色调,简约的家具,冷峻中透着丝丝沉稳和优雅感。在注重装饰效果的同时,设计师用现代的手法和材质还原古典气质,让空间兼具古典与现代的双重审美效果,充分地将生活中的阳光浪漫气息展现在设计当中。

THE ARISTOCRATS CLASSIC NOBLE MANSION

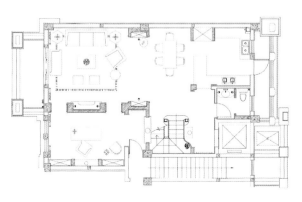
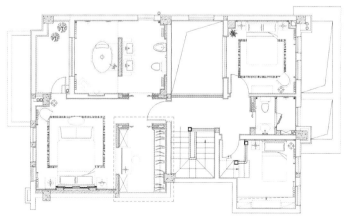

THE ARISTOCRATS CLASSIC NOBLE MANSION

Longgang Fenglinyu

龙岗风临域

Design Company: STEVE.S DESIGN	设计公司：史礼瑞设计师有限公司
Designer: Shi Lirui	设 计 师：史礼瑞
Project Location: Shenzhen	项目地点：深圳

The clear ground colors such as beige, yellow, khaki and coffee are interlacing inside the space. Within the color tone variations, the space becomes more three-dimensional and possesses more emotions. Inside the hall, the marble of three different colors patter collages in the format of diamonds, while the ceiling and wall apply simplified European lines to delineate the boundaries, thus forming the division boundaries of space and surfaces. Apart from upstairs and downstairs functions, the unique spiral staircase forms a wonderful scenery line in the corner. The array and combination format of the spots, lines and surfaces are the new interpretations towards the space by the designer. The marble lines of the public space appear much softer and gentler set off the wallpaper. The handrail of the staircase applies hale black color and the space can extend under the guiding functions of black lines. The transitions between upper and down spaces are greatly connected as a whole.

THE ARISTOCRATS CLASSIC NOBLE MANSION

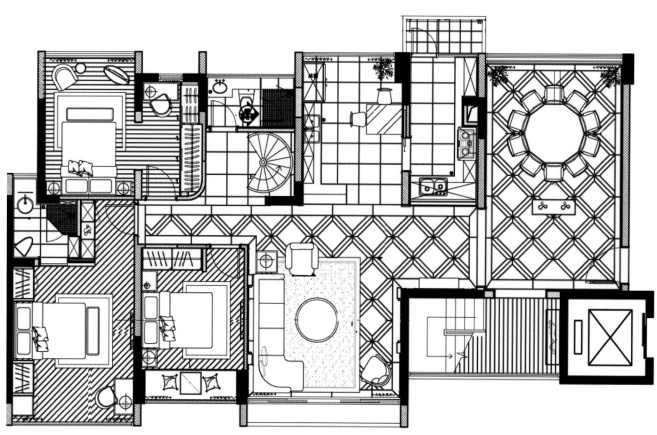

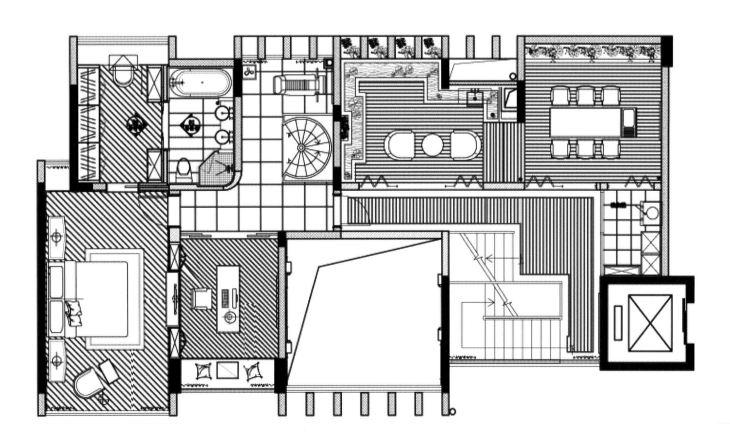

THE ARISTOCRATS CLASSIC NOBLE MANSION

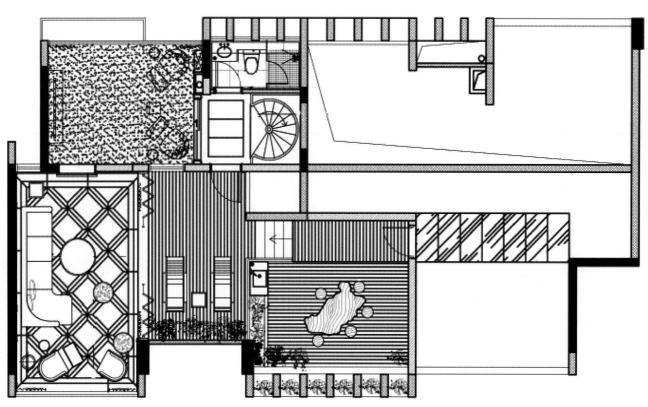

As a private space, the family pictures in many places represent the host's love for home. The grouping cabinets in the study are custom-made for the host by the designer, which go throughout the study. While maximizing the application for the space, the wood veneer pattern and the gold color of the doorknob delicately decorate the space.

The wallpaper of the bedside background applies low-profile and soft gray blue color, echoing the delicacy and gentility of bedding accessories. The whole master bedroom presents romantic atmosphere, finely setting off the nobility and elegance of hostess. With stones as ground and wall surfaces for the wash room, colorful marble pattern naturally creates the space texture.

洁净的米色、黄色、卡其色、咖色这些大地色在空间里错落的交织着，空间在色调的演变中更加立体和富有情感。厅堂里三种不同颜色纹理的大理石以菱形的方式拼接，顶棚、墙面则用简化过的欧式线条勾勒出边界，以作为空间及体面的划分界限。别致的旋转楼梯除了有上、下楼的功能，在角落里也形成了一道靓丽的风景线，点、线、面的排列组合形式是设计师对空间的重新演绎。公共空间的大理石线条在壁纸的衬托下温婉柔和了很多。而楼梯的扶手则用了硬朗的黑色，空间在黑色线条的导视作用下得以延伸，上、下空间的过渡被很好地联系起来。

作为私人空间，多处的全家合影体现了主人对家的热爱，书房的组合书柜也是设计师为主人特别定制的，贯穿整个书房的始末。空间被最大限度利用的同时，木饰面纹理、门把手的金色都使空间被精致地装点起来。

床头背景的壁纸采用了低调柔和的灰蓝色，与床品的细腻和温婉相呼应。整个主卧呈现出浪漫的气息，极好地烘托了女主人的高贵和优雅。卫生间用石材作为地面和墙面，斑斓的大理石纹理自然地描绘出空间的质感。

THE ARISTOCRATS CLASSIC NOBLE MANSION

Rongxin David Country 1

融信大卫城 1

Design Company: Fujian Guoguangyiye Architectural Decoration and Design Engineering Co., Ltd.
Project Examiner: Ye Bin
Designer: Su Wei
Project Location: Fuzhou of Fujian Province
Project Area: 180 m²
Major Materials: Soft Roll, Wallpaper, Jazz White Marble, White Wainscot Board, Gold Tawny Glass

设计公司：福建国广一叶建筑装饰设计工程有限公司
方案审定：叶斌
设 计 师：苏威
项目地点：福建福州
项目面积：180 m²
主要材料：软包、壁纸、雅士白大理石、白色护墙板、金茶镜

This project is oriented to be a high residence of concise European style, changing the complex impressions of European style, replaced with concise lines, white color tones and modern materials to present to the space quality. The open-style layout delineate the magnificent atmosphere of the space, producing some romantic and delicate atmosphere from living room to dining hall. The noble crystal droplights, simplified furnaces and dining chairs of classical carvings present some noble life quality.

本案定义为简欧风格的挑高住宅，一改传统欧式风格的繁复印象，取之以简化的线条、白色为主的基调、现代的材质来表现空间的质感。开放式的格局，把空间的大气氛围勾勒出来，而从客厅过渡到餐厅，则营造出一种浪漫、精致的氛围。华贵的水晶吊灯、简化的壁炉、古典雕花的餐椅等，体现出一种高贵的生活品质。

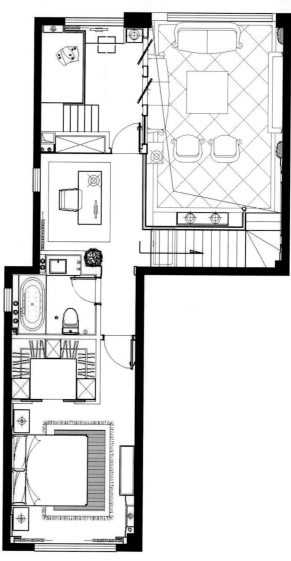

THE ARISTOCRATS CLASSIC NOBLE MANSION

Rongxin David Country 2

融信大卫城 2

Design Company: Fujian Guoguangyiye Architectural Decoration and Design Engineering Co., Ltd.
Project Examiner: Ye Bin
Designer: Ye Meng
Project Location: Fuzhou of Fujian Province
Project Area: 450 m²
Major Materials: Spanish Beige Marble, Light Coffee Web, Silver Dragon Marble, Microlite Bricks, Mirror Surface, Silver Foil, Solid Wood Wall Board, Soft Roll
Photographer: Li Lingyu

设计公司：福建国广一叶建筑装饰设计工程有限公司
方案审定：叶斌
设 计 师：叶猛
项目地点：福建福州
项目面积：450 m²
主要材料：西班牙米黄、浅啡网、银白龙、微晶石砖、镜面、银箔、实木墙板、软包等
摄 影 师：李玲玉

This project applies styles of modern luxury classical style, in plane planning combing the practical functions of villa and circumstances of original house type, integrated with modern life elements, while fully considering the practical quality of the space and performing appropriate space divisions. The space design pursues the transparent feel of the space, with the whole space highlighting the layers feel of Neo-Classical format language with magnificent soft decorations ornaments, strong color combinations, delicate European format elements and abundant variations of light and shadows, being luxurious, dynamic and variable, achieving the noble and magnificent decorative effects.

本案采用现代奢华新古典的风格，平面规划上结合别墅的使用功能与原户型的条件，融入了现代的生活元素，充分考虑空间的实用性，并进行合理的分区。空间设计上则追求空间的通透感，整体通过华丽的软装配饰、浓烈的色彩组合、精美的欧式造型元素及丰富的光影变化，凸显出新古典造型语言的层次感，豪华、动感、多变，从而达到雍容华贵的装饰效果。

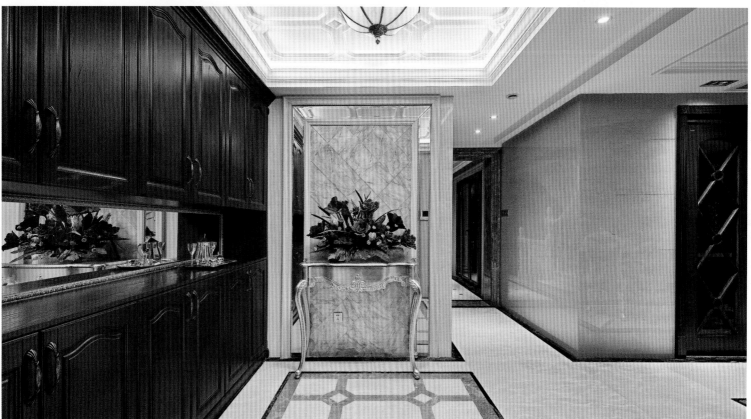

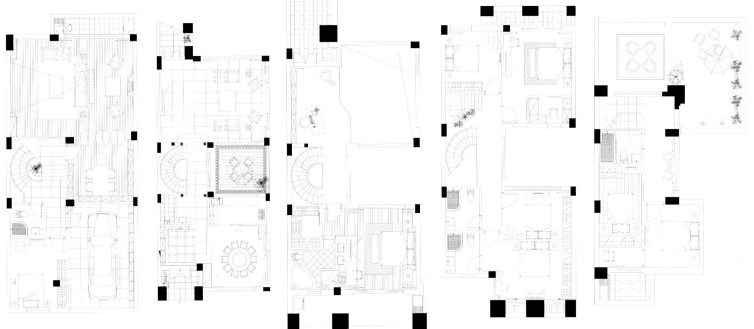

THE ARISTOCRATS CLASSIC NOBLE MANSION

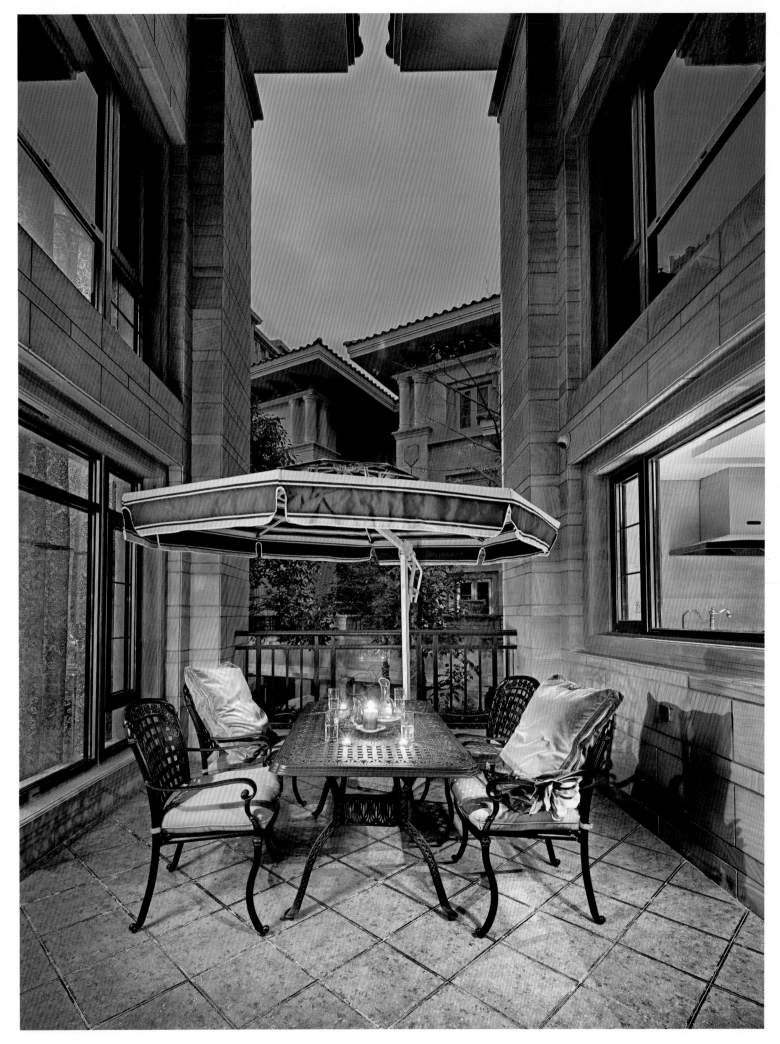

THE ARISTOCRATS CLASSIC NOBLE MANSION

Evergrande Palace Show Flat

恒大华府样板房

Interior Design Company: Hangzhou EHE Interior Design Co., Ltd.	室内设计公司：杭州易和室内设计有限公司
Designer: Ma Hui	设 计 师：马辉
Soft Decoration Furnishing Company: Hangzhou Jishang Design and Decorative Engineering Co., Ltd.	软装陈设公司：杭州极尚装饰设计工程有限公司
Layout Designer: Ge Xulian	陈设设计：葛旭莲
Project Location: Taiyuan of Shanxi Province	项目地点：山西太原
Project Area: 408 m²	项目面积：408 m²

Inside the limited space, the designer organically combines the spot, lines, surfaces with space construction languages, making the whole space be full of luxurious and romantic atmosphere, and making the whole space become more transparent and bright.

Upon entering the space, you can find that the elegant crystal drop-lights send out warm light and shadows, with abstract mural of intensive classical colors on the wall, harmoniously integrating with the whole color tone. The designer ingeniously applies crystal marble ground materials, making the space perform perfect release and extensions. While at the same time, the collocations of different color tones separate the different functional spaces. The carpet of peculiar style, the fabrics sofa of different luster and the dazzling flowers on the tea table make the living room send out noble and dynamic sensations.

Inside the dining hall, the delicate and refined dining accessories and romantic candle sticks clash with each other and the mosaic tiles wall of innovations make the space do not appear monotonous. For the study, the dark coffee color solid wood floor makes the whole space appear sedate and elegant. The tables and chairs perfectly display the Neo-Classical style with "scattered formats and integrated spirits."

设计师在有限的空间里，将点、线、面与空间构成语言有机结合，使整个空间弥漫着奢华与浪漫的气氛，也让整个空间更为通透、明亮。

走进空间，高雅的水晶吊灯洒下暖暖的光影，墙上悬挂着具有强烈古典色彩的抽象壁画，与整体色调和谐统一。设计师在地面上巧妙地运用晶莹剔透的大理石，使空间有了绝美的释放和延伸。

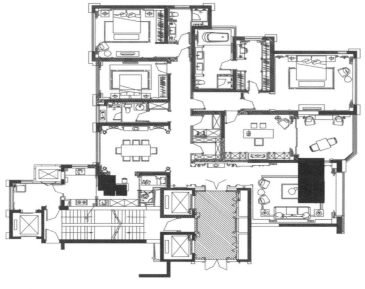

THE ARISTOCRATS CLASSIC NOBLE MANSION

THE ARISTOCRATS CLASSIC NOBLE MANSION

THE ARISTOCRATS CLASSIC NOBLE MANSION

Wooden Love · Shimao Skyscrapers

木质情缘·世茂天城

Design Company: Modern Space Decoration Design	设计公司：摩登空间装饰设计
Designer: Liu Jiayang	设 计 师：刘家洋
Assistant Designer: Liu Weiwei, Zhu Zhongqiang	助理设计：刘伟伟、朱忠强
Project Area: 160 m²	项目面积：160 m²
Major Materials: Solid Wood, Marble, Archaized Brick, Wallpaper	主要材料：实木、大理石、仿古砖、壁纸等
Photographer: Li Lingyu	摄 影 师：李玲玉

This project belongs to be grand plane layout. The designer carries out summaries and recreation towards the plane and facade, producing moving lines of fluent and elegant format and various functional areas which are independent yet connected with each other.

A large amount of wood materials were applied in the space, with free-style and casual American style penetrating throughout the space. The living room is quite spacious and bright, of orderly layout, with antique wood book case and tea table, fabrics sofa of stripes pattern and soft cushion of elegant format with intimate brisk atmosphere. The gold ceiling is collocated with antique little drop-lights and the gold yellow colors possess pleasing comfort feel.

The master bedroom sends out some noble feel in the concise design. The elegant wood bed, nice-looking bed pillars are matched with white soft roll background, with little furnace sending out warm lights. The room does not have ceiling lights, which are set in the ceiling, sending out scattered light inside the space.

THE ARISTOCRATS CLASSIC NOBLE MANSION

The layout of the little bedroom attains leisurely style, the white wood bed is matched with light color bedding accessories and sends out leisurely tastes under the surrounding of brown furniture. For the location near the windowsill, the designer makes use of wood materials to produce a little convex space, with table finely making use of the space.

The wall of child's room is painted blue. Under the background of blue, the white furniture appear much fresher and more natural. The delicate format of furniture and pattern of careful carving send out some noble feel. With ingenious applications, the marble, antique bricks and wallpaper perform some leisurely and luxury feel. While highlighting the space temperament, the design presents the property owner's taste and life attitudes.

本案属于大平层格局，设计师对平面与立面进行了归纳、再造，划分出流畅优美的动线及各个独立而又相互关联的功能分区。

木料被大量运用于空间中，自在、随意的美式风格贯穿始终。客厅十分宽敞、明亮，布置得井井有条，仿古的木质书柜、茶几，条纹图案的布艺沙发、造型优雅的软靠，都带着亲切的明快气息。金色的吊顶配上复古的小吊灯，金黄的色彩，有着宜人的舒适感。

主卧在简约中散发出贵气感，优雅的木床、好看的床柱，配上白色的软包背景，小壁灯透出暖光。房内没有设立顶灯，而是将灯管埋设在吊顶中，散漫的灯光游离于室内。

次卧的布置更富有休闲的情调，白色的木质床配上浅色的床品，在深棕色家具的包围下更加凸显悠闲的气息。靠近窗台的位置，设计师用木料包裹出一个凹形小空间，空间内放置了书桌，很好地利用了空间。

儿童房的墙壁被粉刷成蓝色，白色家具在蓝色背景的衬托下更加清新自然。家具考究的样式，细心雕刻的纹理，散发出贵气感。大理石、仿古砖、壁纸的运用，演绎出了安逸与奢华感。本案在突显空间气质的同时也展现了屋主的品味与生活态度。

Flower Town

香溢花城

Design Company: Fang Kuai Styles	设计公司：方块空间
Designer: Cai Jinsheng	设 计 师：蔡进盛
Project Location: Nanchang of Jiangxi Province	项目地点：江西南昌
Photographer: Deng Jinquan	摄 影 师：邓金泉

This project is narrating a story about time. Time might change people's appearance, and changes things that would not change in people's minds, yet the heart with passion for life and sense of belongings that home brought us would never change.

The arch white TV background wall is like walking into the warm Mediterranean, what the eyes observe are different sceneries.

You stand upstairs observing the sceneries and people enjoying the scenes are standing downstairs looking at you. The cabinets, genuine leather sofa and arch wood door bring you through the long river of time, and it is like getting back to the primitive and pure time.

Among the noisy crowns, is there a person who would hold your hands and look into your eyes? In your own world, you can watch the vicissitudes of life, sighing with emotions, turning around and conducting yourself in society with indifferent attitudes.

THE ARISTOCRATS CLASSIC NOBLE MANSION

THE ARISTOCRATS CLASSIC NOBLE MANSION

THE ARISTOCRATS CLASSIC NOBLE MANSION

本案例是在讲述一个与光阴有关的故事，时间可能会改变人的容貌，改变原来认为不变的东西，但是那颗热爱生活的心以及家所带给我们的归属感是永远不会变的。

拱形的白色电视背景墙，恍若走进了温情的地中海，眼睛看到的是不一样的风景。

你站在楼上看风景，看风景的人站在楼下看你。橱窗、真皮沙发、拱形木门，带你在岁月的长河中穿梭，彷佛回到了质朴纯真的年代。

在那熙熙攘攘的人群中，是否有一个人与你或执手，或相望。在自己的世界里坐看人生的悲欢离合，感叹自己的际遇，转身，淡然处世。

American Charms

美式情怀

Design Company: TOMA International Design
Designer: Hong Bin
Project Location: Fuzhou of Fujian Province
Project Area: 663 m²
Major Materials: PanAmerica Floorboard, Artistic Wall Board, Marmocer Stone, Wanshida Lighting Accessories, Markor Furnishings
Photographer: Zhou Yuedong

设计公司： 唐玛国际设计
设 计 师： 洪斌
项目地点： 福建福州
项目面积： 663 m²
主要材料： 泛美地板、艺术墙板、米洛西石材、万事达灯饰、美克美家家具
摄 影 师： 周跃东

For the project design, for space innovations, the designer perfectly combines the format, materials selection and color layers in space design. With American style as the tone, the design aesthetics integrating classical charms and modern charms create elegant and cozy space atmosphere. The living room retains the large-area ventilation and lighting in the original architecture. The large remote control curtain not only maintains the space privacy, performing great regulations for lights. With natural and fluent design styles, the whole space is delicate, practical, with clear distinctions between active and quiet elements, presenting some healthy and leisurely design concepts.

本案的设计中，设计师在空间创新的同时，也完美地将空间设计中的造型、材质选择、色彩层次结合起来。以美式风格为基调，融合古典情怀与现代的设计美学，营造出典雅舒适的空间氛围。客厅保留了原建筑中大面积的通风与采光，巨大的遥控窗帘不仅保证了空间的私密性，而且还对光线起到了很好的调控作用。整个空间自然流畅的设计风格，别致实用、动静分明，呈现出一种健康、休闲的设计理念。

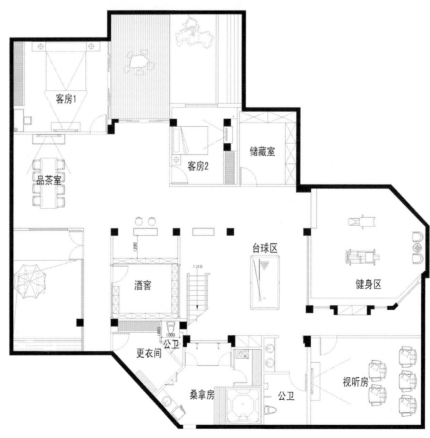

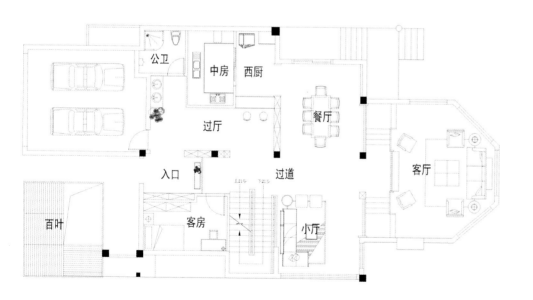

THE ARISTOCRATS CLASSIC NOBLE MANSION

Hereditary Noble

华贵世家

Design Company: Dashu Shangpin-Zhuang Design
Designer: Wang Wei
Soft Decoration Designer: Umbria Professional Soft Decoration Institution
Construction Company: Dashu Construction
Project Location: Changshu of Jiangsu Province
Project Area: 260 m²
Major Materials: Marble, Archaized Brick, Wallpaper, Wood Veneer

设计公司：大墅尚品 - 由伟壮设计
设 计 师：王伟
软装设计：翁布里亚专业软装机构
施工单位：大墅施工
项目地点：江苏苏州
项目面积：260 m²
主要材料：大理石、仿古砖、壁纸、木饰面

Home is the harbor for hearts. This project tries to create some noble and elegant life atmosphere for the inhabitants. This project has fine lighting and ventilations, with proper layout, expressing the space vertical feel and extension feel. The ingenious application of wall tile marble and symmetric design techniques display delicate yet quite warm texture. No matter in what location, you can always get the visual focus. What it produces is not only new visual orientations, but also conveying the elegance and nobility, leisure and romance for the whole residence.

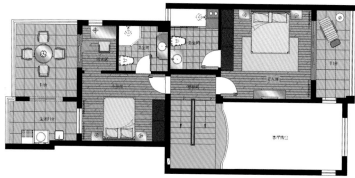

家是心灵的港湾，本案试图为居住者营造出一种高贵典雅的生活氛围。本案的采光、通风较好，布局合理，表达出空间的纵深感及延续感。墙砖大理石的巧妙运用，对称的设计技巧，呈现出精致华丽又不失温馨的质感。无论在哪个方位，站在何处均有视觉的焦点。它产生的不仅仅是视觉新意向，更传达了整个居室的典雅与高贵、惬意与浪漫。

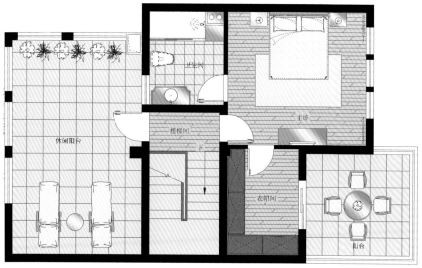

THE ARISTOCRATS CLASSIC NOBLE MANSION

THE ARISTOCRATS CLASSIC NOBLE MANSION

Taipei Huacheng

台北华城

Design Company: Sherwood Design
Designers: Huang Shuheng, Dong Zhongmei, Xu Zongxuan
Soft Decoration Layout Designers: Hu Chunhui, Zhang Hedi, Wu Jialing, Yang Huihan
Project Location: Xinbei City
Project Area: 220 m²
Photographers: Wang Jishou, Zhao Zhicheng
Composer: Cheng Xinchun

设计公司：玄武设计
设 计 师：黄书恒、董仲梅、许棕宣
软装布置：胡春惠、张禾蒂、吴嘉苓、杨惠涵
项目地点：新北
项目面积：220 m²
摄 影 师：王基守、赵志诚
撰　　文：程歆淳

Mix and Match is aesthetic, positive and negative are both elegant.

Walking upwards along the mountain road, you can find a western style villa quietly standing among the green plants, the red blossom trees set off each other and large amount of white walls and the red pitched roof make people feel like being in some other land. Facing the grand mountain views, the designer adjusts the moving lines and applies negative element as positive element, shrinking the entrance of the original facade, just leaving several scattered little windows just for a little light to get inside. The original negative side the building has large area of French windows for scenery for the family members to get together and enjoy the sunny or rainy mountain views in the distance, observing the changing of four seasons. The charms are just like the literary giant of Song Dynasty, Su Dongpo, who said in his poem "Light and heavy luster are both appropriate."

Rebuilding the communication between the outside world and us, and narrating the harmonious home.

This project explores from the highest spot and represents

the ancient charms of British style residence with pitched roof of teak wood, maintaining the high feel of attic and removes the traditional dark impressions. While at the same time, the reading space (as well as the little musical instrument room) is set at the highest spot, as well as the specific practice for "sub-consciousness". This space makes use of large area transparent glass to create fencing, making the visions bright all at once, with the ceiling window gently introducing the sunshine inside.

For designers, the graphics of a nice home is constructed on elasticity. Apart from creating vitality of aesthetics and functions, it also needs to consider the spiritual requirements of each member, allowing people to perform independent work and gather together as well. Out of that, the designer combines geological features with personal preferences, planning the optimum visual spot as a family hall for worshiping the Buddha (as the property owner's exclusive study), containing tranquil, peaceful and natural orientations, with ㄇ format space opposite the grand mountain views. The high wood floor good for storage echoes the dark color wood wall. That is quite in accordance with functional requirements and the tranquil and pleasing atmosphere is quite enough for people's meditation and inspirations would come along endlessly.

Stepping further, you can find the living room neighboring the little garden, which is decorated with several concise paintings. People can read, talk and cook in the environment with no pressure. When it is sunny, people can get to the outdoor space, enjoying the soothing breeze and dancing with the flowers and grass, enjoying the relaxing and pleasing afternoon time.

Things and people are originally natural, and we can enjoy in serenity the rising of clouds.

As the architect's own residence, what this project tries to present is not only the spectacular interpretations on the style by the designer, but also the elements expressing emotions and ideals. The cultural mix-and-match soft decorations design presents the integration of space distance. The interaction of space scales present the designer's experiences in England and the dialogue process fostered by thinking. Through consistent self-questioning and self-persuasion, aesthetics, functions and personal likes are finally integrated as a whole. Among the changing of breeze and clouds, only people and the mountains are constant. The natural and integrating nice posture are like repeatedly narrating the famous lines by poet Wang Wei, "When I feel in the mood, I would walk into the landscape alone. I will walk till the water checks my path, then sit and watch the rising clouds."

混搭即美学　"反""正"皆雅观

沿着山道蜿蜒而上，看见一幢西式的别墅静立于蓊郁之间，粉红的樱树交互掩映，大片的白色墙面和丹色斜屋顶，使人错觉身在异乡。面对眼前堂皇的山景，设计者调整动线、倒反为正，缩小原有正面的入口，只余几扇错落的小窗，允许光线微微射入。建筑原反面处，则以大面落地窗收纳景致，供业主家人闲暇时欢聚，亦可远眺晴雨山色，静观四季跌宕，风韵颇似宋代文豪苏东坡所言："淡妆浓抹总相宜。"

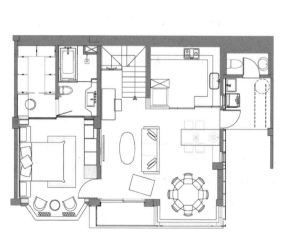 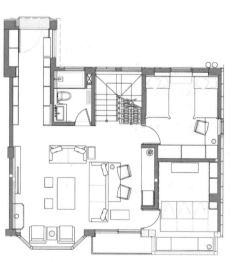 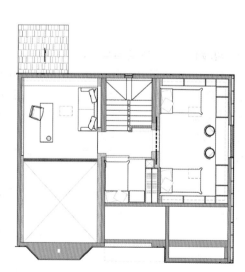

重塑物我交流　谱写和谐家园

本案自最高处开始探索,以柚木装修的斜屋顶,体现了英式住宅的古韵,保持了阁楼的挑高感,消弭了幽暗的传统印象。同时,将阅读空间(亦是小型乐器室)置于建筑最高处的设计,也是对于"潜意识"的具体实践。这个空间利用大面的透明玻璃铺就围栏,使人大开眼界,上开天窗,使日光缓缓洒入。

对设计者而言,美好家园的图像应建构于弹性之上,除了创造美学与机能的生命力,亦需照护每位成员的精神需求,允许人们独立作业、亦可群聚一堂。是故,设计者将地理特性结合个人偏好,将视野最佳之处规划为佛堂(与屋主专用书房),涵纳平心静气、人本自然的取向,将"Π"字型空间正对大幅山景,便于收纳的挑高木地板与深色木墙相映,符合机能需求,其静谧安然的氛围,足供人们静修沉思,灵感因能源源不绝。

前进几步,是紧邻小花园的起居室,只以几幅简单画作点缀墙面,人们得以在无负担的环境里阅读、谈天、洗手作羹汤,天晴时,家人相偕坐到户外,风声和缓、花草共舞,享受轻松惬意的午后时光。

物我本自然　坐看云起时

作为建筑师自宅,台北华城一案欲呈现的,不仅是设计者对风格的独到诠释,更有抒发情志、表彰理想的成份,透过文化混搭的软硬装设计,呈现空间距离的交融,时间尺度的互会,体现了设计者的留英经验,和思维养成的对话过程。透过持续的自问自答、或者自我说服,美学、机能和个人喜好三者终致合流。风云流转之间,唯人和群山各自凝定,其物我自然、万化合一的静美姿态,如将诗人王维的名句反复吟咏:"兴来每独往,胜事空自知。行到水穷处,坐看云起时。"

图书在版编目（CIP）数据

豪门：贵族府邸典藏 / ID Book 图书工作室编 . — 武汉：华中科技大学出版社，2015.3
 ISBN 978-7-5680-0356-8

Ⅰ. ①豪… Ⅱ. ① I … Ⅲ. ①住宅－建筑设计－中国－图集 Ⅳ. ① TU241-64

中国版本图书馆 CIP 数据核字 (2014) 第 191861 号

豪门　贵族府邸典藏　　　　　　　　　　　　　　　　　　　　　　　ID Book 图书工作室　编

出版发行：华中科技大学出版社（中国 · 武汉）
地　　址：武汉市武昌珞喻路 1037 号（邮编：430074）
出 版 人：阮海洪

责任编辑：赵爱华	责任监印：秦　英
责任校对：胡　雪	美术编辑：张　艳

印　　刷：北京佳信达欣艺术印刷有限公司
开　　本：965 mm×1270 mm 1/16
印　　张：21
字　　数：168 千字
版　　次：2015 年 3 月第 1 版第 1 次印刷
定　　价：358 元 (USD 69.99)

投稿热线：(010)64155588-8000　hzjztg@163.com
本书若有印装质量问题，请向出版社营销中心调换
全国免费服务热线：400-6679-118　竭诚为您服务
版权所有　侵权必究